HITTING IN COMBAT

The Brain Science of Training to Win Gunfights

Dustin P. Salomon

and

Innovative Services and Solutions LLC

Silver Point, Tennessee

Thank you for purchasing this book.

If you care about gun safety as much as we do, please visit the website below to access our free video series on safety and fundamental gun handling.

https://www.buildingshooters.com/free

This program contains some of the most important skills anyone who uses firearms will ever learn.

Yet these same skills are still frequently ignored in formal programs of instruction, leading to many avoidable accidents.

Join us in working to make the firearms industry, and society in general, a safer place. Secure your access now.

Copyright © 2021 by Innovative Services and Solutions LLC

All Rights Reserved.

ISBN 13: 978-1-952594-10-6

Imprint: Building Shooters

Library of Congress Control Number: 2021904775

LCCN Imprint Name: Silver Point, Tennessee

Table of Contents

Foreword by the Author 1

Preface: Science in Firearms Training 5

Chapter 1: What It All Means (Definitions) 9

Chapter 2: How It All Works (All About the Brain) 13
    How the brain changes 13
    Fire together, wire together 14
    Fire apart, wire apart 14
    Competition for energy and resources 19
    How the brain learns 20
    A systems view 21
    Input and processing 21
    Storage 22
    Long-term memory structure 23
    Why it matters 24

Chapter 3: Why Non-Visually-Aimed Fire Falls Short 27

Chapter 4: Physics Always Wins 33

Chapter 5: What Is Instinctive? 47

Chapter 6: Historical Performance 51

Chapter 7: Reality-Based Training 59

Chapter 8: Training Period Performance 69

Chapter 9: Market Forces 79

Chapter 10: A Matter of Training 83

    Recommendations and considerations for training 87

Thank You and Join Us! 93

References and Recommended Reading 95

More from Building Shooters 101

NURO: A Brain-Based Analysis of Tactical Training and the Basis of Design for the World's Most Capable Tactical Training System 103

Foreword
by the Author

For readers who are experienced shooters—and very likely instructors—the controversial nature of "point shooting," "kinesthetic shooting," "target focused shooting," and so on (pick your own terminology here) is unlikely to be news. For readers who are new, inexperienced, or prospective shooters, as well as any curious researchers from outside the industry, this book's topic boils down to a single question: *Is it possible to aim in a gunfight?* It might surprise you to learn that this question remains definitively unresolved, without consensus among firearms and tactical trainers.

Intelligent and experienced people sit on all sides of this debate, which should make us wonder how something that seems so simple can be so controversial. One of my objectives with this book is to cut through the emotional noise—on both sides of the issue—and help people see past their own tactical and personal experience-based performance biases.

I aim to produce a scientifically-grounded understanding of the topic, based on what is currently known about how the brain works. As a factual matter, the issue is anything but simple, but instead is incredibly nuanced.

To any new shooters who have somehow made it this far, I am very glad that you are reading this book. The information in it is very important to you. It is *more* important to you than it is to an experienced shooter or to an instructor. Experienced tactical shooters are more likely to fully understand this book, yet you are by far in the best position to *benefit* from it.

Some parts of this book are very technical and assume a substantial level of preexisting knowledge. Readers who are inexperienced in tactical shooting may struggle to grasp some of the concepts being discussed, but the subject matter deserves and requires this level of engagement.

It is my sincere hope that the broader message of this book will still be clear to both those inside and outside the industry: The central message is intended to benefit primarily the student rather than the instructor, though many readers are likely to be instructors. If this describes you, I hope that you will approach this book with an open mind and consider the idea that you may be making an important choice for your students that can have near-permanent impacts on their capability and perhaps even survival.

This choice carries long-term consequences which may forever impact a person's capability to hit the enemy during armed combat. I believe that asking students to make this choice without the requisite knowledge to fully understand it is unfair to both students and instructors, the majority of whom really care about their students. Students deserve choices that are made with informed knowledge and consent—I hope that this book contributes to new consideration regarding ethical and effective approaches to firearms training.

After reading this book, you should fully understand the choice that this issue presents to each person who, through either desire or profession, endeavors to achieve skill-at-arms.

Dustin Salomon

Author and Founder, Building Shooters

Preface:
Science in Firearms Training

There has been some recent criticism of "science" being applied in the firearms training industry. While this may seem at first glance like something from the dark ages, many of those making the critiques have very legitimate points. The word "science" is often used (and not just in the firearms training industry) as a method of deflecting criticism. It is then leveraged as a cudgel to bludgeon anyone who asks questions as being some sort of ignorant, uneducated fool.

This use of the word is unfortunate. There is a lot of danger in using science as a label to provide top cover for an idea because science is not a label: Science is a method of asking questions in pursuit of truth. Real science should not fear questions, because real science is built from asking questions.

Science cannot fear to be wrong, because learning that something is, in fact, demonstrably wrong advances the pursuit of truth and understanding. Real science actively seeks out not only that which is right, but also that which is wrong in pursuit of empirical, definable, repeatable truths. Learning that something is wrong *is doing science.*

The ability of the scientific method to provide answers is only as good as the information that is available at the time of the analysis and the methods of analysis that were used. As a result, "scientific" conclusions are often heavily biased and are limited by factors such as the availability of tools that can provide accurate measurements, the relevance of what we choose to measure, and assumptions about how variables impact each other.

Scientific research and publications are also run by human beings, and like any such system, this means that both study and consensus are subject to influences such as human bias and political pressure. This should not be interpreted as a criticism of science but rather an acknowledgment that human systems (and, in fact, all systems) have limitations.

New information, new methods of measurement, new experiences, and new discoveries can, and frequently do, radically change scientific understanding. Like any other area where people are involved, these developments, and the pace at which they occur, are also heavily influenced by human factors. A perfect example of this is a field of study that is featured quite prominently within this book: neuroplasticity in the adult brain.

Until relatively recently, this was considered by broad scientific consensus to be a debunked, unsupportable theory that was entertained by only quacks and fruitcakes. What a difference new data makes! Thanks to the work of several brave scientific pioneers who risked (and in some cases sacrificed) promising careers by doing actual science rather than dogmatically following the accepted consensus, neuroplasticity in the adult brain is now considered to be established scientific fact. The study of it is considered cutting-edge science and is producing life-changing discoveries in many areas of human development, human performance enhancement, and medicine.

This cautionary example of neuroplasticity illustrates why much of the criticism of science being applied in this industry has some validity:

There is danger in taking one study, one piece of reference material, a single data point, or even long-held consensus and dogmatically making an entire construct based around it. Such an approach can often lead to misapplied, misunderstood research and flat-out wrong conclusions. It is also often used to produce clever marketing ploys for substandard methods or products masquerading as real science, which is applied chiefly as a term to ridicule or discredit anyone who dares to ask questions.

On the flip side of the equation, the reader likely requires no persuasion to acknowledge that there are plenty of hard-headed people in the firearms industry who do things in certain ways and refuse to change for no other reason than "This is how it has always been done." This is where the scientific method can be extraordinarily valuable.

Asking questions such as, "Does it work?" and "Can we do this better?" then applying research findings and using a structured process to answer those questions with the best available data ought not to be controversial. This is particularly true in the tactical firearms training industry where, when it does not work, good people die untimely deaths who otherwise might not have.

Whether we want to call it science or not, everyone should be able to get on board with pursuing what really works. There is tremendous value in trying to understand how things function fundamentally, then using that knowledge to adjust how we do things to achieve better results.

As an industry, we should be building upon the genius and groundbreaking effort of training pioneers such as Sykes, Fairbairn, Applegate, and Cooper. These brilliant men led the way into modern tactical firearms training by asking questions, challenging the status-quo, and doing things that were altogether different than the previously accepted norms to achieve better outcomes.

Questioning their methods is not an insult. We can only truly honor their legacies by following their lead.

Chapter 1:
What It All Means (Definitions)

The first thing we need to do is define some terms that will be used frequently throughout this book. There is frankly a lot of confusion related to this subject matter, and some of the commonly-used verbiage is less than helpful. It is important to sort this all out before we get into the details, so I will start with some definitions and discussion of terms and techniques. I personally do not place much value in clever definitions or word games—it is the concepts and applications that matter. However, for the purposes of discussion between the covers of this book, understanding the material will be much easier if we are all using the same language.

First, let us discuss the common term *point shooting*. There are actually a variety of definitions in common use, and everyone seems to have his or her own nuanced flavor of the day. Here I am going to define pure point shooting as non-visually-aimed-fire. In other words, the sights and body of the weapon are not purposefully used to visually aim the weapon. The shooter's visual focus and attentional focus (two different things) are both directed at the target (threat) throughout the duration of the engagement.

A term the reader may have heard that is often used synonymously with point shooting is *kinesthetic shooting*. According to the *Cambridge*

Dictionary, kinesthetic means "connected with the ability to know where the parts of your body are and how they are moving."

As an experienced firearms instructor once very accurately pointed out during a public discussion on this topic (thank you, Chris S.!), *all* shooting is kinesthetic in nature because all shooting involves use and awareness of the body. From a dictionary definition perspective, the way this terminology is commonly used in the industry is therefore a bit misleading. Nevertheless, when you hear the term *kinesthetic shooting* used, it typically refers to visually non-aimed fire, where visual and attentional focus are placed on the target or threat.

In this book, I use the terms *visually-aimed-fire* and *non-visually-aimed-fire* to refer respectively to (1) the purposeful application of visual skills and attention to an aiming system on the firearm itself and to (2) firing without the use of gun-related visual skills.

There are many different visual aiming techniques and combinations of visual and physical techniques that can be applied to shooting. Most people who shoot, including many instructors, do not fully understand this point; however, it is true.

It is accurate to think of visual aiming skills as a continuum of possibilities that can vary based on the distance to the target, size of the target, and requirement for precision and accuracy. Visually aiming is not limited to the application of a single technique. For a competition-based breakdown of a system of aiming techniques, among other topics, Brian Enos's classic book *Practical Shooting: Beyond Fundamentals* is well worth the time and expense.

It is also important to understand that there is more than one aspect to visual aiming. For our purposes here, we will break it down into two distinct areas: visual focus and attentional focus. Again, the terminology does not matter here other than to provide a common reference within these pages—it is the concepts that are important.

Here are a few simple examples you can try right now from your chair to illustrate and understand this. (*Please note that the purpose of these exercises is to illustrate the dual concepts of visual focus and attentional focus. We are purposefully not discussing techniques or methods for aiming a firearm.*) Wherever you are sitting, find an object about the size of a light switch approximately seven yards or so away. It doesn't matter what the object is: We are just looking for something roughly the size of an index card, several paces distant from you.

First, focus your eyes, as you would focus a camera, on the object. The object should be crystal clear (or as clear as you can make it). Look directly at it so that the object itself is in your focused vision. Then, while keeping your eyes focused (like you might focus a camera) on the object, use your peripheral vision to *pay attention* to what you can see to both the left and the right of you. While it is not commonplace in most people's day-to-day existence, where your eyes are focused and where your attention is focused can, in fact, be two different things.

Next, take your index finger and use it to point at the object, while closing one of your eyes. Look directly at your finger. Focus your eye on your finger, with your finger being clear and the object becoming a blur in the background. This is representative of the most common visual aiming technique taught by firearms instructors to new students, where your finger represents the front sight of the firearm.

Now repeat the repeat the exercise, also with one eye closed, but with a twist. Make a fist instead of using your index finger. Hold your fist horizontally, so that your palm is pointed towards the ground. As you point your fist at the object, focus your eyes on your fist. Your fist should be clear; the object should be a blur in the background.

Align the center of your fist with the object. Visually estimate where the center of your fist is. This is representative of another type of visual aiming technique where the same basic visual skills are used, but much more mental processing and attention are required to get a consistent and accurate alignment. You must manually conduct the "calculations" necessary to assess the center of your fist and align it with the center of the target.

Finally, try that same physical technique again, using your fist with one eye closed. This time, however, focus your eye directly on *the object*. Your fist should be a blur, the object in the distance should be clear. Once again, align the center of your fist with the object, while keeping your *visual* focus (the focus of the lens of your eye) on the object.

Hopefully, these quick, non-shooting-related examples have effectively illustrated the point. There are a broad variety of not only visuomotor skills (what the muscles in your eyes do), but also of attentional focus (information reception and processing) options that can be combined and applied to different activities and skills, including shooting.

Each combination of visual focus method and attentional focus method has different capabilities and limitations that will produce different results for different applications. As should be clear from these examples, things are not nearly so simple as "point shooting" versus aiming.

Chapter 2:
How It All Works (All About the Brain)

Before getting into any discussion of shooting or training, I need to first lay some groundwork by talking a little about how the human brain works. When a person learns how to do something, what he or she is really doing is coding information into the brain, very much like writing code on a computer. The brain is the control system for everything we do. Therefore, as both students and instructors, it is certainly worth asking the question, "How does the brain work?"

This question is more than worth asking—it is critical. If we understand the "terrain" of the brain, we can do a better job navigating it. This allows us to produce better results with greater efficiency in training. With this objective in mind, we are going to discuss some commonly accepted and consistently validated principles of brain function. There are two specific subjects I cover in this chapter, as they are critical to understanding the rest of this book: how the brain changes and how the brain learns.

How the brain changes

We will start with neuroplasticity, or the study of how the brain changes. If you want to dive deeper into this material (and I hope you do), I highly recommend starting with two excellent books, *The Brain that Changes Itself* by Norman Doidge, MD, and *The Mind and the Brain* by Jeffrey Schwartz, MD, and Sharon Begley. You will find a wealth of detailed information that touches upon how our brains work, how they change, and what impacts this can have on people in the real world. Identifying details about these books as well as other references on related topics are contained in the References sections at the end of the present book. Here, I will just summarize three principles that are now well-accepted in the scientific community.

Fire together, wire together

The first principle, popularly known as Hebb's Law, is that neurons which fire together, wire together. As we use our brains repetitively, electrochemical signals travel along the "wires" between neurons. The more that these wires are used, the more efficient the signal transmission becomes—at least to a certain point. This is relatively intuitive. It is easy to understand, especially for shooters, that the more we do something, the better we will get at it.

Fire apart, wire apart

The second principle is that neurons that fire apart, wire apart. This is actually a corollary from the first principle and also an application of Hebb's Law. It means that if two networks are constantly fired separately, they become isolated from each other inside the brain. The result is that any attempt to influence the performance of the skills or skill sequences represented by these networks can become extraordinarily difficult.

Much of the scientific literature on this subject is primarily focused around addressing neurological injuries and illness. However, it also has a great impact on tactical skills training. To understand how this works, visualize an information system that consists of a computer network at a corporate office.

When the network is simply cables, wires, and computers that are hooked directly to the internet, everybody can communicate very easily with everything. The network may be slow, because some employees are streaming videos all day at their desks, but it is also easy to connect and communicate with everybody, using any service such as social media, chatrooms, etc.

The corporate management recognizes that productivity is down and that critical software of the company is running very slowly, so they bring in some IT specialists to get things squared away. The new IT department begins to implement some changes. They establish firewalls, block streaming, blacklist social media sites, prioritize traffic for critical company software, improve authentication and security protocols for network access, etc.

After several weeks of implementation, these changes make the network much faster and much more reliable for the specific functions that are authorized by IT and company management. However, everything else has become much more difficult, if not impossible, to connect to. Running a high bandwidth processing function on the company's newest cloud application? Performance should be excellent. Want to spend the afternoon streaming last night's music video awards? This might present more of a challenge, including special permission from, and network access configuration changes by, someone in IT.

While this is clearly just an analogy, a similar process occurs in the brain. When circuits and networks are repetitively used for highly specific purposes, they undergo changes that make them far more efficient. Unfortunately, the same physical changes that make them efficient also limit their ability to either communicate with other networks or be repurposed for other uses.

This effect is responsible for many of the undesirable responses that develop in, particularly, experienced range shooters when skills become well-learned in siloed isolation. There are many examples in shooting of highly isolated skills performance that can develop because of the common ways we train both on the range and in dryfire.

I was once running an instructor development class. One of the students was a multi-decade law enforcement firearms instructor. As part of the course, we worked on a method to integrate visual response into live-fire training. I told the student that I was going to press the buzzer on a shot timer but that he was not to react until he saw the correct visual stimulus.

What happened when the buzzer went off? He drew his pistol and fired two rounds into the target, then re-holstered. He immediately looked sheepish. I looked at him, we had a chuckle, then we discussed the drill again. He clearly understood the exercise. I pressed the buzzer. What happened? He drew and fired two rounds into the target. This response to that stimulus was so ingrained in him that he was repeatedly unable to control it, even in a "no stress" setting.

In another instance, I was attending a patrol rifle instructor development program as a student during a time when I was still filling an operational role as my primary job. We were conducting a basic trigger reset drill, such as one might conduct with a student during his or her very first time pressing the trigger.

The drill was run at about the four-yard line, just far enough away to prevent the 5.56 muzzle blast from ripping the targets to shreds. The exercise was simple enough. Press the trigger once; hold it to the rear. When instructed, release the trigger enough to reset it and then prep it to fire again. There was no time constraint. There was no stress. It was simply an exercise to demonstrate the mechanical aspects of the trigger and trigger finger management.

I could not do it. I literally *could not* prevent my trigger finger from immediately resetting the trigger and prepping to fire again. My partner for the drill tried to coach me through it. Finally, the course instructors came over and tried, to no avail. Within that proximity to what I was viewing as a simulated threat, I was quite simply unable to break into or change the performance for trigger management that I had accessed and applied by default based on context and spatial relationships.

About an hour later, the class was in the prone position at 25 yards, learning about various methods for zeroing the rifle. Instantly, I had no problem applying that method of trigger manipulation, as one might use for target shooting or precision marksmanship. After the zeroing exercise we had a break, and the instructors let me go back to close range.

With a different mental framework and, presumably, different neural network for skill performance now in place, I had no trouble at all performing the reset drill. However, when the skill network applied was based in a mental framework for close quarters combat, I was physically unable to force my finger to do what I wanted it to do, even in a no-stress, administrative setting.

For a great illustration of this, ask somebody who only ever shoots IPSC (where two rounds are typically fired at each target) to fire three rounds per target. You are likely to see that they will struggle to perform something different than the specific trained sequence of two sequential trigger presses. I jokingly call this IPSC finger; however, it is a very real thing. Ironically, the more that a shooter trains to shoot *only* IPSC style drills, no matter how good they get, the more difficult it can often become for them to do something different than the specific drills that they practice.

The personal examples given here are anecdotal; however, they are examples of issues that commonly manifest in training environments, especially among experienced shooters who have invested significant time and effort into skills training. The fundamental principle of neuroplasticity—fire apart, wire apart—is a big part of the reason that these types of events occur. An actual physical separation and isolation of brain circuits happens when we train skills repetitively in siloed, isolated settings. This happens because of *how* the brain works mechanically.

While the following explanation is *very* simplified, completely ignoring synapses as well as the electrochemical methods of signal transmission that are actually used in the brain (please read *at least* the two books referenced above should you be interested in the mechanics), you can understand generally how this separation happens and how it can impact tactical skill development for combative applications, by visualizing a simple electrical analogy.

Imagine a new skill being developed during training by connecting two electrodes (neurons) with a bare metal wire (called an axon in the brain) running through a crowded cable tray. As signals are repeatedly sent through this wire, the brain realizes that this circuit is important.

The new circuit (skill) is used a lot. Therefore, the brain decides to insulate it. The brain covers the wire with insulation (a fatty substance called myelin). This both protects the signal carried within the wire and improves the efficiency and speed of the signal.

As more of the wire's surface becomes coated with insulation the signal becomes better protected, more efficient, and transmission speed becomes faster. This is what happens when a skill becomes well-learned.

The wires between the neurons get insulated. (Fire together, wire together.)

Insulation, however, works both ways. When the circuit is *not* insulated, it is relatively easy to connect something else to it. All it might take is connecting another wire, with a different signal, in somewhere along the length of the existing circuit's wiring. This non-insulated circuit might not be energy efficient or fast; however, it is still relatively easy to connect other circuits with.

However, after almost everything in the circuit is coated with insulation, this is no longer the case. After the insulation is in place, the circuit and the signal it carries are stabilized and protected. The signal transmission becomes faster and more efficient. The circuit is also isolated from signals that are sent by the rest of the brain. The insulation also acts to *limit other signals getting in.*

The signal transmission within the circuit is efficient; however, the circuit itself is also now much more difficult to modify or network with other circuits and functions. If circuits are made separately, and well-practiced in isolation, they can eventually become stabilized and individually insulated, separate from one another. They, quite literally, wire apart.

Competition for energy and resources

The third principle of how the brain changes is that it is designed to be energy efficient. All brain functions compete for energy and other resources, including neural real estate. They constantly "fight" over what parts of the brain are dedicated to performing specific tasks. The more that something is done, the more important the brain thinks it is. In many cases, this means that more of the brain's real estate will be assigned to complete that task.

Mechanically, the brain can be thought of as a group of biological, electrochemical circuits. Because it is a physical system, the fundamental principles of physics apply to its function. For example, the reader is certainly familiar with the elementary principle that energy and electricity always want to travel the easiest path with the least resistance.

The brain is structured to survive by conserving its resources. It always seeks to conserve energy and will generally use the *least* amount of energy possible in any given situation. The energy it does use, of course, follows the laws of physics and travels the easiest, most efficient pathway to its destination.

This would seem to make us completely mechanical creatures, purely at the mercy of physics and chemistry. However, while the brain's physical and chemical processes have great influence, we also have the capacity to apply what Schwartz calls *mental force*. Under certain circumstances we have the capability of *intentionally* spending energy and *consciously* making our brains perform.

These three principles are extremely powerful when understood and applied. Combined with a basic understanding of how the brain processes and retains new information, they can give us a treasure trove of insight into firearms training methods and human performance.

How the brain learns

This is the subject of my 2016 book, *Building Shooters*. If you want to learn more about this and see how it can be applied to improve training design and delivery methods for tactical applications, or if you want to see the scientific research behind this discussion, *Building Shooters* is the place to start. Here, I will again provide only a brief overview for those not familiar with the territory.

A systems view

It is helpful to think of the brain as an information system rather than just a biological organ. There are, in fact, many similarities between how the brain works and how advanced computing and information systems work.

The reason for this is that the human brain *is* an information system—the most advanced yet discovered in the universe. It is used as a model that drives much of the development in advanced information processing and high-performance computing.

What follows is a very simple model of the brain. It is not intended to be a biologically accurate depiction. It *is* intended to help the reader understand how our brains process and learn new information—as currently understood by modern science.

Input and processing

All new information is received by one of the five senses. It is important to understand that the senses receive a massive amount of data. In fact, they receive so much that if people were aware of all of it, we would be unable to function effectively. To keep us from becoming overwhelmed, the brain has a filter. This prevents most of the information received by the senses from ever reaching the brain's processing and storage centers.

If information manages to get through the filter, it then enters into a space called short-term memory. You can think of this like the random-access memory (RAM) in a computer. It stores a small amount of information that can be worked with and used temporarily. However, it does not retain anything for the long term.

Storage

If information is going to be stored, it first has to be *physically moved* from the short-term system into one, or both, of two long-term memory storage systems. This process is not automatic, and it does not happen instantly. Storing information in the brain requires complex chemical reactions, the synthesis of proteins, and generation of new physical structures. Most of the research says that it takes at least twenty-four hours before new information can be transferred out of short-term memory and stored in long-term memory.

Other important things to know about short-term memory are (1) it is very small as compared to the rest of the brain's storage capacity, (2) it is also not very good at protecting information. Data within it can easily become corrupted. Before new information can be effectively transferred into long-term memory, it must be free of corruption while it is in short-term memory.

Corruption happens when similar, but different, data are used by the brain within relatively short time spans. When this happens, the different information sets can interfere with one another. This leaves no clear set of data for the brain to recognize or store in long-term memory. When this corruption occurs, it is called *interference* in scientific circles.

A good example of this from shooting is handgun grip. Crush grip, high-thumb Weaver grip, and competition/combat grip with thumbs forward are all relatively mainstream "legitimate" methods of holding a handgun for combat shooting. They all involve the same tool, the same muscle groups, the same body parts, and are used for exactly the same purpose. However, they are also all different and rely on different mechanics to control recoil and deliver combat-accurate rounds on target.

These different grip techniques cannot be performed at the same time. Each technique, done well, can work adequately for combative applications. However, a mish-mash hybrid of these techniques is unlikely to produce good results. Using a different grip every time the gun is drawn is even worse and will certainly produce poor results.

Unfortunately, many people are taught and asked to try multiple grip options on their first exposure to firearms. As a result, they never learn a consistent grip and therefore will always struggle to achieve a basic level of skill performance. If you visit any public (or police) handgun range, you will almost certainly see this in action.

Before information is transferred to, and stored, in long-term memory, the short-term memory system must recognize that the information is important. Otherwise, it is likely to simply be wiped out of the short-term space so it can be freed up for something new.

To review, once information makes it through the brain's filter, survives without interference in short-term memory, *and* is recognized as important enough for storage, then the brain can finally go through the process of moving it into one, or both, of the long-term memory storage systems. Remember that this process takes *at least* 24 hours, starting from the time the information is first stabilized inside short-term memory.

Long-term memory structure

Especially for firearms and tactical applications, there are some very important things to know about long-term memory. As I have previously mentioned, there are two separate systems. These are, in some cases, redundant with respect to what information is stored in them (the same information can be stored in both places). However, they have completely different functions and uses.

The first long-term system is commonly called declarative memory. This system only stores information that is consciously accessed. Cognitively-driven, intentional access is required to use the information.

The second long-term system is called procedural memory. This system stores information that is *unconsciously* accessed. Of particular importance for tactical training applications is the fact that *this is the only memory system that is reliably accessible when a person is under stress.*

This means that if a skill is going to be performed in highly stressful environments, *that skill must be stored in the procedural memory system.* It is not enough that the person understands the skill, can describe the skill, can perform the skill, or even can do the skill very, very well. A skill must be stored in long-term procedural memory if it is going to be performed in combat.

Why it matters

This is where our current discussion converges with the previous chapter. In order to effectively learn combative shooting (i.e., learn *and retain* so that the skills automatically happen in combat) it is not enough to understand, or even to be able to skillfully perform, a technique.

To engage a threat with a firearm in combat conditions, a combination of the visual focus method, attentional focus method, and physical motor skills for weapons operation all need to be linked together and stored in long-term memory. For these things to happen under stress, they also need to be coded into an energy efficient network that is located—very specifically—in the brain's procedural memory system.

It would be nice if things, neurologically, were as simple to teach as "point the gun and either aim or don't aim." However, this is simply not representative of reality. Visually aiming can involve dozens of different combinations of visual skills and methods of attentional focus. Furthermore, each of these combinations of visual skills, attentional focus, and physical skills may represent an altogether different technique, each with its own unique network in the brain!

It does not take an advanced degree in neuroscience to understand that each of these different neural networks will have its own unique properties. It is also intuitive to understand that any one of these combinations must be well trained before it can be used in combat. It is further intuitive that when you need to (1) put a lot of mental effort (mental force) into doing something and (2) make actual physical changes to the body's status, such as moving muscles in the eye, that calories are burned and energy is expended in the brain.

In fact, this is the physiological basis for one of the arguments consistently made by those who advocate teaching non-visually-aimed-fire as a primary technique. Visually-aimed-fire, by whatever method, is relatively difficult. It is not "natural." The body, brain, and eyes do not "instinctively" do the things necessary for visually-aimed-fire. Therefore, the non-visually-aimed-fire theory says, we should not bother doing them. Instead, we should train like we will fight: by only doing what the *untrained* body and brain instinctively will do.

Before moving to the next chapter, consider the following question: If we are going to strictly limit ourselves (and our students) to what the *untrained* human body and brain will do, are we really benefitting from our training?

Chapter 3:
Why Non-Visually-Aimed Fire Falls Short

Now that we have a handle on definitions, visual skills, and know how the brain works, let's start talking about shooting. Before we begin, I want to make one point absolutely crystal clear: Non-visually-aimed-fire *does not fail* as a combat shooting technique!

In fact, this method of shooting is a very important skill with critical real-world applications. Because this is so important, I will repeat it for emphasis. Non-visually-aimed-fire is a completely valid (and very important) combative shooting skill that has been taught as a specific shooting method and successfully used to win gunfights for nearly 100 years.

The basic concept behind non-visually-aimed-fire is that your body can "naturally" point at an object in the distance and get pretty darn close. Ardent advocates for teaching non-visually-aimed-fire as the primary shooting technique will suggest that you can validate this right now by picking an object in the distance, looking at it, and then pointing your finger at it. Your finger will most likely be naturally aligned with the object. If you can do this with a finger, the theory is that surely you can do it with a firearm.

Since our bodies can already point, we can just more or less eliminate the need for performing the relatively difficult, energy expending, and potentially time-consuming (depending on the technique used and quality of visual alignment needed) visuomotor skills and attentional focus methods necessary to aim visually.

This is especially valuable because a person's natural inclination will be to look at a person who is trying to kill them, not some little piece of metal atop a pistol slide. It is further a valid point that, if you have decided to shoot somebody, you are presumably already looking at that person. Otherwise, how would you know that deadly force is necessary, and how did you make the decision?

If you change what you are looking at (look at and/or pay attention to the sights), then an extra expenditure of energy is required. There are also stress-related chemical responses in the body that some research indicates may make these specific visual tasks more difficult than they normally would be. For example, the lens in the eye can be flattened, reducing clarity in close-range vision.

The approach of using non-visually-aimed-fire as the primary shooting technique is based on the idea that, if we rely on the kinesthetic alignment abilities of our body's vestibular system, we make the task of hitting a target in combat much easier and more aligned (pardon the pun) with how our bodies naturally work under stress.

This all makes good sense, which is part of why this method of training was developed and became popular. Another significant contributing factor to its popularity is that it is comparatively easy to teach. And it works, too: The basic techniques are combat proven over decades in both law enforcement and military settings. Unfortunately, there is a catch. It is more accurate to say that non-visually-aimed-fire works—right up until it stops working. Then there is a real, serious problem.

The issue fundamentally lies, not with the method of non-visually-aimed-fire, but rather with a combination of its limited capabilities and the energy expenditure requirement involved in its performance, when this is compared against the energy expenditure requirements of shooting techniques that use visually-aimed-fire. This is fundamentally a training and learning problem, not a technique problem. It also leads us to the choice about which this book is written: a choice about how to train.

There is an awful lot packed into the previous paragraph, and we will spend most of the rest of this book breaking it down. The first statement is that non-visually-aimed-fire has limited applicability. Put simply, if you do not aim, with a visual sighting method, you are unlikely to hit what you are shooting at very consistently when there is any requirement for accuracy or precision.

This is something even the most ardent supporters and proponents of non-visually-aimed-fire training freely admit, including, yes, Applegate, Fairbairn, and Sykes. Consistent accuracy and precision can only be achieved by using a visual aiming method. There is no way around it.

A few years prior to publication of this book, I attended Bill Rogers' Shooting School. The school has been operating since the late 1970s, and its test has become something of an industry performance benchmark. (If you are unfamiliar with Rogers' school and test, you can look it up here: https://www.rogersshootingschool.com).

There are three levels of qualification: Basic, Intermediate, and Advanced. Rogers says, quite matter-of-factly, that no one has ever even *qualified* on his target system using non-visually-aimed techniques.

This apparently includes a number of prominent and vocal (though unnamed) advocates for the methodology.

It is worth noting that a number of the points fired on the test are actually at static, torso-sized steel plates. A shooter can miss a full 50% of the pop-up 8-inch plates (many of which occur predictably at about 8 yards) and still qualify.

An intermediate rating on the test can be achieved without even shooting at anything past 12 yards. Furthermore, the most difficult shot in the entire course is an 8-inch plate at under 25 yards. In the context of this discussion, the takeaway is that long-distance, precision shooting requirements such as those required for bullseye target shooting are not an issue.

Is the Rogers' test representative of combat? No. However, it is noteworthy because it requires not only a reasonable level of accuracy, but also that the accuracy be delivered in compressed time frames. Target shooting techniques will not work.

A more relevant question is whether the skills necessary to pass the Rogers test are representative of combat-related skills. It is true that range shooting is not gunfighting. It is equally true that if we cannot perform a skill successfully on the range, we will almost certainly fail to do it in a gunfight.

From a tactical perspective, the problem (or problems) that we want to solve, whether as armed civilians or armed professionals, should inform and decide our equipment, techniques, and training. It is the year 2021 as of this writing. Therefore, I am going to assume that one of the "problems" that everyone who carries a firearm wants the ability to solve is an "active shooter" threat. Before moving on, let me say that I despise this term with a passion for a variety of reasons that I will not get into here. Therefore, I will not use it again in this book. Instead, I will use the term *indiscriminate mass murderer*. (Note that indiscriminate mass murder attacks are by no means the only "civilian" tactical problem that demands the capability for accuracy. However, in the interests of time and space, I will ignore the many others here.)

I will be blunt: If you (or your students) do not have the capability of delivering visually-aimed-fire in combat conditions, then you are not likely to successfully solve this problem. You may well solve a close-range ambush type attack, the scenario for which non-visually-aimed-fire training was intended. However, you will most likely fail to effectively stop an indiscriminate mass murderer, whether you have a firearm with you or not.

(Note: As with many such issues, there are shades of gray in this discussion. Many indiscriminate mass murder attacks are stopped by unarmed intervention. In some cases, they have been stopped as soon as an intervention of some sort occurs. Sometimes, the murderer takes his or her own life. In the context of this discussion, "effective" response involves the successful application of lethal force, where the force itself immediately incapacitates the perpetrator and prevents further murder or attempted murder from occurring.)

Without using the sighting system of the weapon, it is extremely difficult to consistently hit the vital areas of a human being in combative conditions at any distance exceeding a few yards. Let us also not forget that a basic requirement for indiscriminate mass murder is a target-rich environment for the would-be killer—full of possible victims. For the good guys, this means that missing the intended target, even in an ultimately successful engagement, can very possibly result in both civil liability and criminal charges (for you!).

To see but one real-world example, consider the shooting that occurred December 29, 2019 at the West Freeway Church of

Christ in White Settlement, Texas. As of the publication of this book, you can see pictures and a link to video of the event here: https://www.youtube.com/watch?v=i8TI21qcJsk.

Church security team member Jack Wilson successfully stopped an intended mass murder event with a visually-aimed surgical shot that dropped the attacker. It is also noteworthy that, due to his position relative to the attacker, he appears to have fired over the heads of other churchgoers sitting in the pews. The real world, especially for civilians, does not often facilitate "down range" environments as they exist on a shooting range.

This shot cannot be predictably made without using visually-aimed-fire. Even if this were possible, the training time and effort required would be extraordinary, far exceeding that required to achieve application of visually-aimed-fire under combat conditions.

It is also noteworthy when watching this video that Wilson was not the only armed member of the congregation. He was simply the only person who was able to fire a round. Watch the video again. The first person shot appears to be another member of the church security team, who was killed while fumbling unsuccessfully for his weapon—and likely killed *because* he had a weapon—the use of which he was clearly unskilled in.

What is important to understand at this point is that if you (or your students) cannot apply visually-aimed-fire in combat conditions, then there are a host of eminently solvable real-world tactical problems that you are unlikely to have the capability to address. This is true regardless of whether you carry a firearm or not and irrespective of what type of firearm you carry.

Chapter 4:
Physics Always Wins

The human brain is incredibly complex and, even now, not terribly well understood by even the most advanced levels of neuroscience. What is thought to be currently known about the actual processes that govern the material discussed in this chapter involves incredibly nuanced combinations of subjects such as quantum mechanics and highly complex chemistry. I will not discuss these things here partially because they are far out of my depth (and likely yours as well—if not, please contact me as I would love to pick your brain), but also because the actual mechanics involved are not particularly relevant.

Consider the example of a dropped glass. Calculating the full forces, mechanics, and exact final outcome of dropping a drinking glass on a concrete floor—down to the resting location, size, and shape of every glass shard—would be so complex as to be virtually impossible to accomplish, even with supercomputers. However, one certainly does not need a Ph.D. in mechanical engineering and access to a bank of supercomputers to understand or predict accurately that a glass will, in fact, shatter if it is dropped on a concrete floor.

The same concept is at work here. Modern science has observed enough about the brain to have reached a thus-far unchallenged consensus that it adheres to the principle of energy conservation.

Therefore, while the descriptions in this chapter are admittedly not presented in an academically scientific way, the *concepts* are nevertheless accurate to the best of my current understanding. You can find the relevant supporting literature among the references listed at the end of this book.

If you have read any other Building Shooters books or articles, you are already aware that I like developing and using models as tools to help evaluate and understand complex processes and environments. In the context of models intended for training design applications, it is of course important that the model conform with the relevant scientific understanding and explain what has been previously observed. It is also critically important for trainers to have models that enable them to *predict* the impacts that their training will have on the student as well as the outcomes this will produce in the field.

This is the intent of the learning model presented in my book *Building Shooters*. It is also the intent of the model contained in this chapter, which is intended to further both the instructor's understanding of how the science discussed in this book applies to the topic of aiming as well as his or her ability to understand and predict the impacts and outcomes *with respect to students* that will result from various methods of training. This chapter is not a science lesson. However, the model used here is strongly supported by current scientific understanding of neurological function.

It should be unarguable that there is a legitimate tactical requirement to perform visually-aimed-fire-level accuracy on the street—even for civilians. Whether this applies to any single individual at any given point in time is, well, subject to Mr. Murphy and his long history of "looking out" for those who fail to prepare. Does this type of accuracy requirement exist in every shooting? No. However, it *does* exist. To assert otherwise is simply false.

Unless you already know *exactly* what your students' (or your own) eventual shooting(s) are going to look like, as well as when and where they will happen (in which case you are clearly in the wrong profession), you and your students need both combatively relevant visually-aimed-fire (as opposed to target shooting type marksmanship) and non-visually-aimed-fire shooting skills tucked away in the quiver that we call the procedural memory system.

Without both, it is impossible to have a fully functional armed skillset that is relevant to the tactical situation that exists in 2021. This is true for civilians. It is also true for armed professionals, though the tactical needs of each may be somewhat different.

As we discussed conceptually in Chapter 2, there are a variety of different visual skills and attentional focus combinations that can be applied to visually aim a firearm. Each one of these combinations comprises its own unique method of skill performance. Each one also corresponds to its own neural circuitry in the brain.

The use of the term "circuitry" here is important—because that is literally what it is. As we discussed in the last chapter, the brain, at least mechanically speaking, is a vast array of biological electrochemical circuitry. Like any other physical system, it is governed by physics. With respect to electricity and energy, they always follow the easiest path.

When we train a skill to the point of becoming well-learned, we create a neurological circuit (or flow path for energy) in the brain. Hebb's Law tells us that neurons that fire together repetitively eventually wire together, thus creating purpose-built neural circuitry via the training process.

It is critical to understand the following point: When we train and thereby build effective skills, physical changes occur within the brain. An actual physical circuit is created. As shooters, and instructors of shooters, these concepts should be relatively intuitive and non-controversial. We can all see this play out in the development of our students, as well as the development of our own skills.

Here is a summary of what has been covered in this chapter so far:

1. A functional combative skillset with firearms requires the ability to apply *both* visually-aimed-fire and non-visually-aimed-fire (point shooting). These should be applied as appropriate to solve the tactical problem at hand.
2. The brain is made up of electrical circuitry.
3. The brain's circuitry follows the laws of physics.
4. Developing a well-trained skill means creating a purpose-driven circuit in the brain.

Got it? Keep all of this in mind.

For the sake of simplicity, I am going to pretend that there are only two technique options with respect to "aiming" a firearm in the following discussion. We already know this is not true. There are many options. However, this assumption will make the concepts in this chapter easier to understand. In the context of our discussion, there are the two options in play.

1. "Flash-aimed-fire": Sight picture and sight alignment are rapidly assumed. Both visual and attentional focus are rapidly shifted to the physical sighting system that is attached to the weapon.
2. Purely "kinesthetic" or "target focused" shooting: Visual and attentional focus are completely driven to the intended target. The shooter's vestibular system is completely in charge of hitting the target.

If these are the two possible techniques, then we have the option to create the following two circuits in the brain:

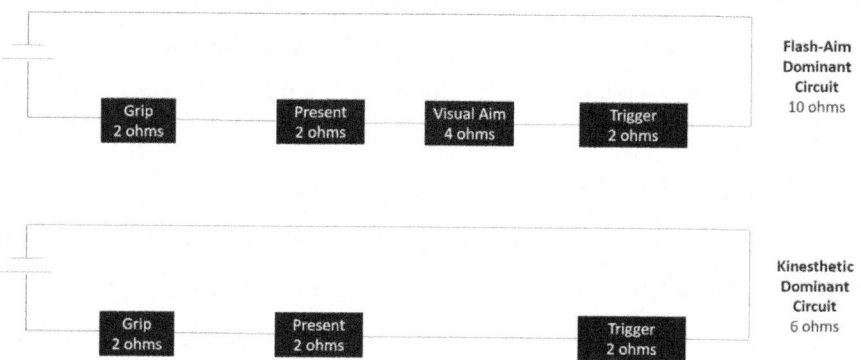

Notice that each subcomponent of the skill (each itself a sub-circuit) is presented as a measure of resistance within the complete skill circuit. For example, gripping the gun is shown as a two-ohm resistor and so-on. (Note that this is intended to be neither a technical shooting nor an electrical engineering guide.)

You may have also noticed that each circuit is labeled as *Dominant*. What does this mean? It means that only one of these circuits (the dominant one—only one of them can actually have this status at any given time) is likely to be performed under stress.

Two questions should immediately come to mind: (1) Which one will be dominant? (2) But I thought you said we needed both skills?

A rudimentary understanding of circuits and electricity should make it relatively easy to understand which technique will be dominant, assuming they are both well-trained and stored in the long-term unconscious access memory system. Electricity travels the easiest path. Therefore, if all else is equal, the one with the least electrical resistance is the most likely to be used when under stress.

There is also a second component that contributes to the dominance of a neural network: context. You can undoubtedly think of an example where you personally have done something that was less than the most efficient option when rushed or under a moderate level of stress. In retrospect, you can look back and see a more efficient method. However, in the moment, you did what was at the forefront of your mind based on the situation and environment.

Readers who have spent significant time in the tactical world are likely familiar with the concept of training scars. These are inefficient, ineffective actions (often completely nonsensical, such as stopping to pick up spent shell casings while under fire or handing a weapon back to an attacker after successfully wrestling it away) that can occur in combat situations.

The action does not need to make sense cognitively. The brain just needs to find it to be the most relevant action associated with the context. To learn more about context and firearms training, please see my earlier book *On Training: Volume 2* and also my book *NURO* (an excerpt is included at the back of this book), which contains a brain-based analysis of tactical training.

It is *theoretically* possible to split two dominant skills and use both of them in different situations through applied context. However, since we are talking about applying deadly force with a firearm for both competing circuits, and since you want to keep both circuits "active," the contexts are probably going to overlap significantly. This makes even the theoretical potential for separating different handgun shooting methods with context muddled and limited, even in the best-case scenario.

Rather than attempting to develop separate and competing techniques and methods, what is preferable to create is something more like a dominant circuit for shooting that has a "bypass" in it. When other techniques besides the dominant one are also present, access to them is governed by something analogous to a switch in an electrical circuit. While there is only one *dominant* technique, other options may still be available.

The neural circuitry in our example can be represented by one of the two diagrams below. They represent the two circuitry options that shooters (and trainers) could seek to create through their training programs:

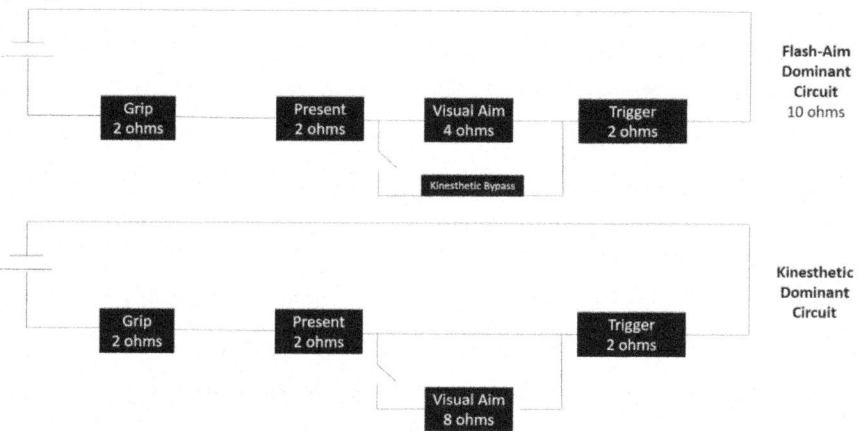

Notice in the "Flash-Aim Circuit" that the kinesthetic shooting option equates to shooting without the visual skill requirements of flash-aimed-fire. The shooter is just not doing those visual skills or providing that attentional focus, though they are performing the remainder of the physical skill sequence. This is similar to the concept of drawing or presenting a weapon without pressing the trigger.

You may also have noticed that the "Visual Aim" sub-skill is rated at 4 ohms in the dominant flash-aimed circuitry and at 8 ohms in the dominant kinesthetic circuit—within the flash-aimed-fire bypass. Why is it more difficult to visually aim in the dominant kinesthetic circuit? Hebb's Law in action.

Creating a dominant flash-aimed-fire circuit involves high repetition of flash-aimed-fire techniques during training. This "wires together" the flash-aimed-fire circuitry, both providing more insulation on the wires (myelin on axons) and improving connectivity at the synapses. All of this strengthens the neural connections which results in less resistance and faster signal flow as the skill improves.

Repeated practice and improved proficiency not only make the brain more likely to identify that particular skill performance method as dominant, they, literally, produce changes in the brain's architecture that make it easier for the signal to flow. *This means that shooters who train for flash-aimed-fire as the dominant engagement technique expend much less effort and energy to visually aim when shooting than shooters who train primarily to shoot using kinesthetic shooting do.*

You may have also noticed that, in both circuits, there is a "bypass switch" located between presentation and trigger manipulation. (Please ignore minutia of technical shooting mechanics for the various techniques here—this is a conceptual example, simplified for purpose.) This "switch" represents an option that exists within each circuit.

During most low-stress training evolutions, closing this switch (thereby sending energy through the bypass) isn't too difficult in either circuit (assuming of course that the bypass has been constructed). We can get signals through the bypass in *both* circuits during low stress periods because we have the capability of applying "mental force" to our own brains when our cognitive functions are available to us.

Cognitive control permits us to mentally drive alterations in the performance of neural circuitry by consciously sending signals to places and through routes that they would not go without our focused efforts. I highly recommend reading Schwartz and Begley's book *The Mind and the Brain* to learn more about the mechanics and scientific evidence for this concept.

Because our brains appear to have this ability, it is not too much of an issue on the range to use either aiming method—on demand—regardless of how a person has trained. However, in high stress environments, things change significantly. Recall that stress triggers chemical releases in the brain that act as a "switch" to the unconscious (procedural) memory system.

Further recall that the procedural memory system is for *unconscious* assess only and therefore is far less influenced by mental force as well as prone to producing automaticity of response. Whatever circuitry exists in this specific memory system is probably going to get used "as is" when the chips are down.

This is why teaching (or practicing) non-visually-aimed-fire shooting in such a way that it becomes the dominant response is such a big deal. This is also why *how* early training is conducted with respect to the methods used to align the gun with the intended target represents such an important choice in a shooter's developmental process. Look at the following diagrams to see what happens when we "close the switch" to engage the respective bypass in each of the two possible "aiming" circuit options.

Flash-Aim Dominant Circuit
Kinesthetic circuit applied under stress
6 ohms *vs* 10 ohms

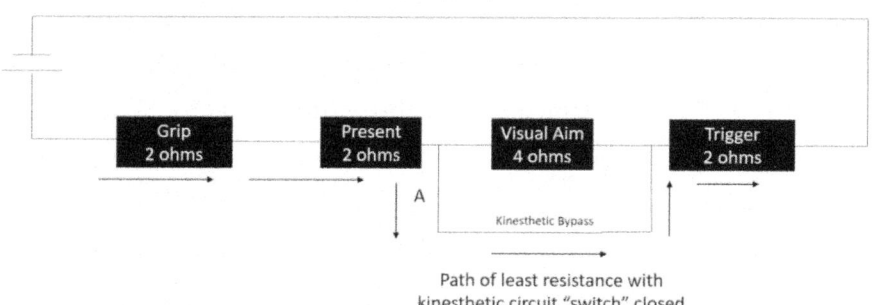

Path of least resistance with
kinesthetic circuit "switch" closed

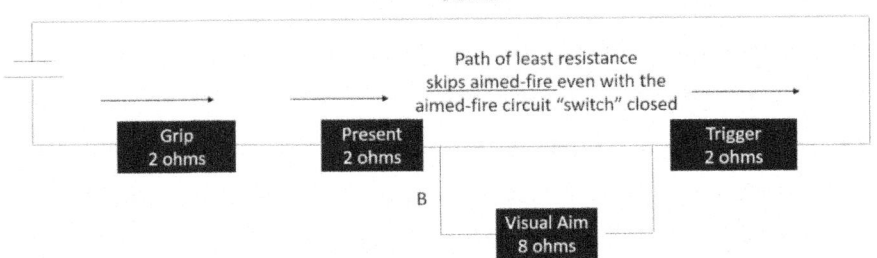

Kinesthetic Dominant
Circuit with aimed-fire circuit applied under stress
6 Ohms

The illustrations demonstrate the issue. If a shooter's dominant circuitry is based on flash-aimed-fire, shooting with non-aimed-fire methods uses *less* energy than the dominant circuit does.

Even though the dominant engagement method is visually-aimed, non-visually-aimed-fire shooting techniques (if trained appropriately) can still be applied with minimal mental force and/or an automatically trained response to stimuli such as obscured or broken visual sighting systems. This can be accomplished even when the unconscious memory system is in control. Once the bypass circuit is activated, it now becomes the path of least resistance.

The reverse situation, however, is *not* true. *If the dominant neural circuitry is based on non-visually-aimed-fire shooting methods, the shooter cannot apply aimed fire under stress.* In this case, the "bypass" resistance value is much higher than that of the dominant circuit.

Not only must another skill (visually aiming) be performed when the "switch" is closed, the visual aiming skill also *adds* resistance to the circuit. As a result, it is virtually impossible for signals to make it through the visually-aimed-fire circuitry under stress. This is not because non-visually-aimed-fire is wrong or bad (it's not). It is just that physics always wins.

There is yet another component to this discussion, leading to the choice around which this book is written. Once neural circuitry is built into the procedural memory system, it is *very* hard to undo. It is often a "first cut is the deepest" sort of scenario, particularly when there is a relative difference in the energy required for performance.

Once non-visually-aimed-fire has become coded into the brain as the dominant shooting method, it is extraordinarily difficult to change it. In practice, this has the effect of not just making it so that a person cannot visually aim in a gunfight but also of making it so that a person *cannot learn* to visually aim in a gunfight. Due to the brain circuitry created, this becomes a more-or-less, permanent state when the procedural memory system is in play. Certainly, this will be the case for the vast majority of students.

I mentioned in the Foreword that, with respect to the long-standing controversy about whether or not it is possible to aim in a gunfight, both sides of the argument were right. That may not have made sense when you read it. Hopefully it does now.

The physics that governs neurological function (as well as the experience of myself and many, many others) indicates that a person who trains early on in combatively relevant visually-aimed-fire and who develops this visual skill as part of the dominant shooting technique will reflexively visually aim under stress. However, a person who trains so that non-visually-aimed-fire becomes the dominant skill will not. For this student, it will quite literally be impossible.

Because there is so much controversy around these topics, please allow me to reiterate here that *both* visually-aimed-fire and non-visually-aimed-fire are important parts of a complete gunfighting ability, and we should stop fighting about them as an industry. Using good tactics is not about dogma. It is about applying the right techniques and methods at the right times, on the right terrain, for the right reasons to accomplish an objective.

This concept also extends to training techniques and methods. By all means, teach and practice non-visually-aimed-fire. In my opinion, you should be doing so. Even with a dominant technique that includes visual aiming skills, *you still must build the circuitry* before you can access and apply non-visually-aimed-shooting skills under stress. This means you still need to learn and practice them if you want a complete, functional combative skillset.

There are several reasons that developing these skills is necessary. I do not believe, however, that this has anything to do with speed. Combat-related visually-aimed-fire is simply not slower than non-visually-aimed-fire. It involves some additional mental processing and the performance of non-standard visual skills (i.e., things people do not normally do as they interact naturally with their environments) during the same time frame as physical skills are performed, but it is not slower in my experience, nor in the experience of others who are far more skilled than I am.

All methods of visually-aimed fire are skill sequences that involve visual interface with, and visual processing of, a reference point from or on the firearm. There are a variety of different ways this can happen that range from red dot sights (RDS) to using the entire silhouette profile of the weapon. I will not discuss specific shooting techniques in this book. What is extremely important to understand, however, is that, regardless of what method of visual aiming is used, there is some type of purposeful visual interface with the weapon.

This involves physical visual skills, attentional focus, and information signal processing that become a component of the shooting skill. This is great for shooting performance; however, we must also recognize that it is entirely possible for this specific visual interface to be unavailable during the moment of truth.

Bad stuff tends to happen in situations where people need to shoot other people. In fact, it is typically a prerequisite. In very harsh settings, batteries can die. Optics can fail. Sights can break off guns. Visual acuity can be occluded from any number of things ranging from ambient light levels, lighting effects, mud, blood, chunks of skin and other soft tissue, glass, dust, dirt, etc.

Even the most perfectly refined and combatively relevant visually-aimed-fire shooting skillset can run into a brick wall if a critical component of that skill's performance is not present or available. To borrow one common phrasing (without commenting on its technical accuracy), you may have the rest of your life to *find the front sight and press the trigger*. This may be a short time indeed if there is no front sight to find. In that case, the trigger may never get pressed, and the problem may never get solved if finding the sight visually is an absolute requirement for skill performance.

Non-visually-aimed-fire can also be extremely important in close-quarters situations, especially at contact distances. Weapons retention (preventing an adversary from physically taking one's gun away) and other defensive tactics and combat considerations can become of primary importance here. It may not even be possible to extend the gun towards the threat in many situations, thereby precluding the use of visually-aimed-fire altogether.

Non-visually-aimed-fire is a set of skills. This means that it can be (and, in my view, should be) trained, practiced, and improved on, like any other shooting skill. If you assume you can shoot effectively without visually aiming in a gunfight simply because you are good at combat-relevant visually-aimed-fire, think again. The applicable skills will be relatively easy to master and will naturally play off the existing motor mechanics you have developed. Please do recognize, however, that non-visually-aimed skill application does use somewhat different circuitry in the brain.

While these shooting techniques are important, I encourage you to think very long and hard about the limitations you may place on yourself or your students if non-visually-aimed-fire becomes the dominant method of shooting. It certainly has the potential to do long-term damage to performance potential and possibly even negatively impact survivability.

It is my very strong suggestion that you consider building a dominant circuit of a combatively applicable, visually-aimed-fire technique first, *then* introduce appropriate non-visually-aimed-fire shooting techniques that are applied for the right tactical reasons, in the right settings, and are associated with the right contexts.

By applying appropriate training design, you can make the physics work in your favor and construct a purpose-built non-aimed-fire bypass within the dominant aimed-fire circuitry. Build capability through your training, not limitations!

Chapter 5:
What Is Instinctive?

As I have said repetitively in this book, this topic is not new. It is one of the most contentious topics throughout the entire firearms and tactical training industry. As with any controversial topic that sticks around for a while, there are several consistent discussion points and arguments used. In the interests of attempting to further understanding of the relevant issues, I am going to walk through the most common ones in the next set of chapters, point by point.

My goal is to apply the knowledge we have learned in the first four chapters of this book and to take as much of an analytical, dispassionate view of the subject matter as possible. After completing this book, is it my hope that you will better understand these complex issues and be able to make a fully informed choice about how you choose to train and/or teach.

One of the most intriguing components of the ongoing debate in the shooting community about visually-aimed-fire versus non-visually-aimed-fire is the assertion that humans naturally and instinctively can point at something. Often this is "proved" to students by asking them to pick an item or spot in the distance and then point their index finger at it. When the student's finger quickly arrives on the point they chose, the declaration is that this has demonstrated the instinctive nature of pointing. But has it?

In this exercise, is not the student achieving a visual alignment with the desired target, verified with visual attention, visual focus (or both) on the finger itself? If not, how can he or she verify that the pointing worked and that the finger is aligned with the chosen spot target? Is not the student, during this exercise, actually *visually aiming*? Food for thought.

There is a further point to understand on this topic. Having read a significant amount of scientific literature relating to brain function and human performance development, I believe it is a virtual certainty that the ability to point and, indeed, the body's ability to interact with the world around it is learned, rather than being truly instinctive such as involuntary functions like heartbeat and digestion.

We are not born with the innate ability to successfully do much of anything. From the moment we take our first breath, life becomes a learning process. We develop an understanding of how our body interacts with the world around us through constant trial and error as we navigate through our surroundings.

From the perspective of human development, there appears to be little doubt that relational functions and awareness of our body's place in the environment that surrounds it are, in fact, learned experientially. Furthermore, the dynamic nature of sensory and motor map locations in the brain seems to strongly indicate that these types of relational awareness functions are, rather than being innately "hardwired," constantly being "trained" (*exactly* what part of the brain is used for specific things is not consistent from day to day, even for the same person).

With respect to the subject of this book, this is important because the reason we are successfully able to point at objects is not due to some genetic coding that programs us to do so. Rather, this ability is because we *continuously train* ourselves to point with adequate precision and accuracy as we go about living our lives and interacting with objects in our immediate environment.

Consider the following question: If we effectively train ourselves to point effectively—without putting forth any effort at all—what might we be able to train ourselves to do with just a little bit of focus and effort?

There are people in the training and shooting industry who make the statement to students or other shooters that it is not possible to learn to use, or to see, a gun's sights under stress. In fact, I recall receiving an email from a reader several years ago asking me how much time I had spent reviewing the scientific literature proving that visually aiming a firearm under stress was impossible. I found it an odd question because I personally have visually aimed a firearm in stressful settings and know many others who have done the same. It is somewhat strange to assert that a skill many people perform as a matter of course is "impossible."

I encourage you *not* to take my word on this. For but three references, I suggest that you review Sergeant Major (ret.) Kyle Lamb's book *Stay in the Fight! Warrior's Guide to the Combat Pistol*, Sergeant Major (ret.) Patrick McNamara's book *TAPS: Tactical Application of Practical Shooting*, and also listen to Michael Seeklander's interview with Massad Ayoob on the American Warrior Show in 2017 (at the time of this writing, this is available at the following link: https://americanwarriorshow.libsyn.com/one-on-one-with-massad-ayoob-self-defense-expert-and-author-of-in-the-gravest-extreme).

If you are unfamiliar with any of these folks, Kyle Lamb and Patrick McNamara are two highly-experienced special operations combat veterans with decades of experience in the most elite military units. They both have extensive personal experience not only performing in combat, but also preparing others for combat and leading them into it. There are no mysteries and no conjectures in their world about what is and is not possible in combat and other stressful environments. Massad Ayoob is perhaps the industry's most prolific author and researcher. He has interviewed countless gunfight participants throughout his decades-long career and is a highly skilled shooter and recognized expert on human performance in his own right. He knows what questions to ask and how to ask them.

There is generally plenty of room for discussion and for differing opinions on all kinds of topics. However, the statement—made as a fact—that aiming in a gunfight cannot be done is grade A baloney. This is not true. Something that many people do regularly cannot, by definition, be impossible. When using the appropriate techniques, visually-aimed-fire in combat situations is absolutely possible. It simply requires a measure of training—the same way that pointing does.

Chapter 6:
Historical Performance

Another common discussion point related to this topic centers around the dismal records of police hit rates—when they are trained to use visually-aimed-fire techniques. This is a very important topic to discuss because it is absolutely true and, in fact, was one of the driving forces behind the development of non-visually-aimed-fire as a formal shooting system. It also speaks to the crux of the issue that has led me to write this book.

Police (and much of the military) are trained to pass qualification standards structured around some combination of traditional marksmanship skills or "bullseye" shooting. Sometimes, particularly for law enforcement, shooting standards are also influenced by—in my view—a somewhat misinterpreted analysis of data relating to standard job or mission performance requirements.

For example, suppose that an agency's historical data indicates that 85% percent of rounds are fired from distances of 1-3 yards during on-the-job performance in real-world shootings. Because of this, it is often required that 85% percent of the rounds fired during that agency's training and qualification occur between 1-3 yards.

It is undeniable that these approaches (especially the use of traditional marksmanship skills) usually produce extraordinarily poor performance results during real gunfights. There is even an argument to be made that aggregate gunfight performance, at least in some jurisdictions, might improve if the police had no training at all (note that I am neither making this argument nor suggesting no training for police).

It is critical to understand two points here regarding skills and training methods. First, as we have already discussed in this book, the aiming and marksmanship skills taught and applied in support of passing more traditional qualification courses are, quite literally, different skills than those related to combative shooting. (If you are curious about the history of this topic, at a minimum I recommend picking up a copy of *Shooting to Live* by W.E. Fairbairn and E.A. Sykes.)

Back in the early 1900s, non-visually-aimed-fire with handguns was first formally developed into a training system by a couple of British officers (Sykes and Fairbairn) to support policing in Shanghai. At the time, virtually all formal training with handguns was single-handed, bullseye-style shooting. There was no gun handling involved, simply deliberate application of the fundamentals of marksmanship.

This type of shooting is incredibly difficult and requires enormous levels of discipline and patience to master. It also has virtually no practical application in the real world, or at least has not since formal pistol duels over matters of honor became a deserved relic of the past. Bullseye-style marksmanship typically does not even involve deployment of the weapon from any sort of carry or position of readiness. Most events start with the shooter already in an extended shooting position.

Personnel trained only in this manner, unsurprisingly, fared very poorly in close-quarters gunfights with criminals, mostly occurring in low light, on the crowded streets of Shanghai. With no mechanism or method to present a weapon, and a practiced shooting skill sequence that begins with a very deliberate and precise visual focus on a tiny black sliver of metal atop the pistol slide (invisible in even moderate darkness), traditional marksmanship training failed miserably. Its main contribution was getting police officers shot dead in the street.

While the techniques and methods for shooting skill that are taught to modern law enforcement have developed quite a bit in the past one-hundred years, unfortunately, it remains a fact that most of the firearms-related skills that are *learned* (as differentiated from what an instructor may have taught) and performed during qualification and training today are still unrelated to the requirements of the real world. This is especially true with respect to visual skills.

Depending on the shooter, determining this from observation may require a more finely-tuned-eye than the average person possesses. It may even require some in-depth, probing questions to see and understand the differences. However, most shooters trained specifically for shooting standard law enforcement qualifications use different grip, different recoil management, different trigger management and manipulation, different attentional focus, and different visuomotor skills than are appropriate (or sometimes even useful) in combat or other stressful settings where the sympathetic nervous system is in control. Literally, the physical skills, related neurological pathways, and attentional focus used are different.

Today's techniques look more relevant than the one-handed target shooting of the early twentieth century. However, on average, the skills used during training and qualification are still different than those that will work optimally in a gunfight.

The second, and equally important, issue in play is that the instructional systems and teaching methods that are predominantly used to "train" people (including law enforcement) in firearms and other combative skills are fundamentally misaligned with human cognitive architecture, as we saw in Chapter 2. Said another way, the training methods don't work very well. They simply do not match up with how the brain receives and learns information.

Again, some historical perspective here is valuable. Fairbairn and Sykes are generally considered legends in the shooting and training community—and rightly so. They are among the most influential combat skills trainers of the past century. It is, therefore, worth considering what they actually did and why it was so ingenious.

If you invest the time to read *Shooting to Live*, you might be surprised to see the details of the specific program of instruction that is outlined within its pages. You will see that it is measured in hours, not days, consists of mostly dry weapons handling, and that far fewer live rounds are fired during the entire process (at least as outlined in the book) than are fired during a single course of qualification fire by modern military, law enforcement, security officers, and even on civilian concealed carry tests.

Like many—probably even most—trainers in institutional settings, they had limited access to their students, limited resources, and an unforgiving work environment that required a constant churn of bodies to fill roles, skilled or not. This is how armed workforces operate, particularly outside of elite units. People are not there to train. They are there to do a job. Training may be part of the job, but it is usually only there to facilitate putting bodies into billets.

What we have previously discussed about how the brain works and learns in the first half of this book should make clear the reality. The *training methods* developed and used in Shanghai, as presented in

Shooting to Live, did not work all that well. They could not. The human brain does not learn that way.

Fairbairn and Sykes did not develop a method of building skilled, or even competent, combative shooters. What they did was develop and formalize the performance of techniques that allowed unskilled people to kill effectively with handguns at close range. This was accomplished by applying learning transfer to gross motor skills that all people are already trained to do—specifically reaching/pointing at an object in front of them. The only thing new students needed to actually *learn* was how the tool worked. Fairbairn and Sykes also determined, through trial and error, the bare minimum resources necessary to be reasonably sure they had achieved this result with an average student.

If this commentary comes across as critical, it should not. This was pure, tactical brilliance. When faced with a problem that they could not solve (teaching a large group of people to shoot effectively in combat with no time or resources), they recognized that they actually did not need to solve that problem. They just needed to get students to kill with handguns at very close range and let the gun do the work. They also realized that the shooting skills required to kill at close range with a gun are quite minimal—and proved this quite effectively on the streets of Shanghai.

During the Second World War, this knowledge was then passed on to others, including an American officer named Rex Applegate—another deserved legend in the industry. Applegate's assignment was to develop combative skills training, including with firearms. This was then taught to specialized operatives who were sent to work behind enemy lines.

After significant research, Applegate modified the techniques developed by Fairbairn and Sykes to improve accuracy by extending the weapon at eye level. He then began teaching these techniques to students before they went into combat environments.

As one would expect, the training timelines were greatly compressed and resources were limited. In some cases, trainees (mostly intelligence operatives) were only given a few hours in total to learn and practice combative skills before being sent off behind enemy lines.

Applegate's challenge was similar to the one faced by Fairbairn and Sykes in Shanghai. He did not need to build effective shooters with most of his students—that wasn't important. He needed to teach people who were actually on their way to war (ready or not) to kill in combat without the time or resources necessary to make his students *learn* anything new at a procedural level.

Unsurprisingly, the same approach employed by Fairbairn and Sykes was effective. Leveraging skills that the students were already trained to do minimized the need to teach them much of anything new. Students only needed to slightly modify (apply learning transfer to) existing skills that are already practiced on a daily basis in order to get the job of killing at very close range done effectively.

This historical perspective is enlightening for several reasons. First, it throws into sharp relief the historic challenges faced by earlier generations—challenges that frankly dwarf anything faced today, even in today's law enforcement climate. Fairbairn, Sykes, Applegate, and their contemporaries were faced with the problem of training people for combat with time and resources that provided exactly *zero* chance of making the students learn anything new and put it into procedural memory. Learning was neurologically impossible.

Their solutions were pure genius. These men figured out how to teach people to kill—very effectively—without learning anything new—simply by applying existing, well-trained skills in a novel manner.

The second reason that this historical context is enlightening is that it also highlights one of the two fundamental conditions (the other being a lack of illuminated weapons sights) that drove the development of non-visually-aimed-fire as a dominant shooting technique in training. The limitations of the training timeline, literally, made it *scientifically impossible for the students to learn*. Given this limitation, the trainers were forced to teach things the students already knew so they could be modified for a specific purpose.

Fast forward to today. Unfortunately, some things have not changed much. Even if the techniques and methods being learned (again, as opposed to taught) by most law enforcement were appropriate for the intended application (which they are not), the *way* most law enforcement are trained, despite access to training time and resources that would have made Fairbairn and Sykes green with envy, simply is not capable of producing effective learning.

Consider this statement: Most shooters who attend formal training (including most law enforcement officers) practice techniques that are known not to work well in a gunfight (regardless of what the instructor may have taught them) in such a way that they could not learn them in any case.

It is against this sad reality that historical police performance in gunfights must be evaluated. Arguing against visually-aimed-fire based on this data is equivalent to arguing against issuing body armor to deploying troops based on analyzing results from a returning unit that was only issued IIA soft armor (no plates) and rarely wore it on patrol: "Body armor didn't prevent a single casualty on that deployment. Let's stop issuing it."

If you wouldn't make that argument, don't make the argument against visually-aimed-fire shooting techniques, at least not based on historical police gunfight data. It simply is not relevant.

There is also another point to be made here. The systemic problem *with respect to shooting technique* (judgment is another issue) is not that police officers don't fire their weapons. Rather, it is that they historically miss most of the rounds that they fire because they shoot without being able to apply the traditional "bullseye" marksmanship skills they have learned to perform in support of passing qualification. In other words, the consistently poor performance of police in gunfights over many decades is—wait for it—the real-world results of non-visually-aimed fire.

Let me be clear (again). I am not arguing against non-visually-aimed-fire as a valuable technique. It is an important, combat-proven skill with critical real-world applications. I am also not suggesting that non-visually-aimed-fire cannot be trained, nor that its performance cannot be greatly improved with some minimal modification of existing physical skills. This can be done, and the training to accomplish these improvements *should* be done when appropriate.

I am simply pointing out the irony that much of the police shooting data that is often used to argue against teaching visually-aimed-fire techniques actually shows the results of non-visually-aimed-fire performed under high stress. The results are from an unpracticed application of it, to be sure, but there it is.

Chapter 7:
Reality-Based Training

Before getting into the meat of this chapter, let me be crystal clear to avoid confusion: Force-on-force, in addition to other experiential training, is a *critical* component of a functional training system. Please do note, however, that poorly-designed, scenario-based-training can be harmful to student development. If you design, deliver, or participate in this type of training, please see my previous book *On Training: Volume 1*. It includes a chapter about one of the ways this can happen. I also encourage you to check out at least two other excellent books on the subject, *Training at the Speed of Life*, by Kenneth Murray, and *A Scientific Approach to Reality Based Training*, by Jeff Quail and Dr. Terry Wollert.

For engagements within the effective range of the simulated munitions (airsoft, UTM®, Simunition®, etc.), if your shooting methods and techniques do not work in force-on-force training, then they won't work for real. Period. If your skillset doesn't work successfully in well-designed, scenario-based training, then you need to go back to the drawing board, regardless of how successful you may be on the range.

However, the opposite of this statement is *not* true. Just because something works well in scenario-based training does not mean that it will work in the real world. Like anything else, scenario-based training has limitations. It is therefore important for trainers to understand what they are, lest we unknowingly lead our students down the well-intentioned path to operational failure.

Since this book is all about training visual skills, we will stick just to that specific subject matter in this chapter. The following considerations are *only* about the impacts and relationships of force-on-force to firearms aiming techniques, at least within the context of this chapter.

It is important to understand that force-on-force training, at least with today's technology and tools, contains multiple "false" *visual* aiming systems. Although these aiming systems exist when using simulated munitions, they do *not* exist when firing live rounds.

In the interests of time, space, and readability, I will not dive into great detail on the mechanics here, but there are some additional neurological principles that are important. Our visuomotor, visual processing, spatial awareness, self-awareness, and visually-directed skill performance abilities are all significantly impacted by two general principles of human function.

First, we are goal-oriented. The mind and body are predisposed to function in accordance with a defined and recognized objective. We are then driven by our intent to accomplish it at both the cognitive and precognitive levels.

Second, the mind and body are largely error-based in terms of performance and performance development. Success in task accomplishment is achieved through correcting identified errors during repeated performance attempts.

This means that a significant percentage of the brain signals involved in both orienting information and in making physical changes or movements (both conscious and unconscious) are the result of continuous feedback that is received at multiple locations throughout the brain's data stream from multiple sensory systems. This, of course, includes the visual system and various other types of signal processing within the brain.

We detect errors and then compensate for them to achieve our goals. This is not something that is learned or trained (like pointing). This is how our bodies and brains work at a fundamental level. It is *how* we receive the data that our brains use to learn.

For a crude example, think of a guy peeing in a toilet while standing up. He might hit the rim initially, depending on what angle it's coming out at, but he's going to get the rest of it in the bowl once he can see where it's impacting and adjust accordingly.

There is no training required for this (though female readers who have ever shared a bathroom with a male are probably engaging in some sort of internal monologue at the moment). There is simply a requirement for the knowledge of where the stuff is *supposed* to end up and the ability to receive the visual feedback from where the stream is impacting.

Unfortunately, when utilizing non-visually-aimed-shooting techniques, limitations in most force-on-force training tools are exploited by these fundamental neurological functions to provide two redundant, "fake" aiming systems. Specifically, these are the flight path of the projectile and the impact of the projectile (especially when using marking cartridges).

For the benefit of those readers who have not done so, when you shoot Simunition®, UTM ®, airsoft, paintball, etc. and you are *looking at the simulated threat* (visually-threat-focused shooting) while you fire, you can see not only the impacts of the rounds on and/or around your intended target, but you can also see the flight path of those rounds in first your peripheral and then central vision. This is especially useful if you are firing rapidly (as most people do in force-on-force scenarios) and multiple projectiles are in the air at the same time. This creates something of a "stream" of visible rounds in the air.

Note that these streams of airborne projectiles are *not* things you need to be trained to look for or to compensate for. These are visual stimuli that your eye will naturally see (because they are both in your field of vision and are visual representations of change/motion, which both your eye structures and neural structures, as well as the related processing systems detect very well) *and* that your mind and body will use and adjust for to accomplish the objective. It is also important to recognize that this goal-focused adjustment to accomplish the intended task starts occurring at a *precognitive level* in your brain's processing systems.

Want to test this out? Use one of the mentioned simulator training tools if you have access. Alternatively go pick up a cheap airsoft gun. You could even use a squirt gun, garden hose, or squirt bottle and get the idea. Stand at 7-10 yards or as close to that as you can get using what you have and shoot as rapidly as you can from the hip to get as tight a group as you can within a time period of 3-5 rounds or so. You can try this with a SIRT or other laser pistol too, but you won't get the "stream," only the impact.

If you can do so safely on a live-fire range, try the same thing live-fire, then compare your groups. If you have low-light training ability, backlight the target and use a cardboard, plastic, or thick cloth flap with some stand-off to keep light from coming out of the impact holes so you can't see the impacts on the target, and then try it. You can also throw a dark-colored shirt on the target to help obscure bullet impacts (full light or low light).

(Live-Fire Safety Note: Don't try any of this live-fire if you're not a highly-skilled and experienced shooter with access to professional-level facilities. Don't do this if you're not 110% confident you can do it safely. Make absolutely sure every round you shoot is hitting the berm/bullet trap or impacting in the surveyed SDZ (Surface Danger Zone). Take an unloaded weapon and take a knee to see what the angles towards the impact area look like from hip level and consider that you will have poor elevation control of the muzzle from the hip, especially when firing live rounds.

If you needed me to tell you any of this, I highly recommend that you do not try this live-fire without in-person guidance and oversight from a professional instructor who really knows combative shooting on a professional-quality range. If you're not absolutely certain you can do this with 100% safety and 100% bullet containment, then don't do it at all.)

After this experiment, hopefully you can see that "threat focused" shooting with these simulation tools is *not* non-visually-aimed-fire. It is actually visually-aimed-fire using visual sighting systems that will never exist in the real world.

But that's not all: Not only do today's most common force-on-force technologies provide not just one, but *two* "fake" aiming systems that start working at a precognitive level, they also fail to provide an adequate representation of two of the most significant impediments to effective firearms use: recoil and the concussive effect created by centerfire cartridges when fired without a suppressor.

Both are relevant, but I will just focus on recoil here. The simple fact is that, even ignoring visuomotor skills and aiming methods, fundamental shooting mechanics (grip, body position, trigger management, etc.), which perform adequately or even well with simulated munitions, may fail catastrophically when used with live rounds. This is especially true when using a major caliber service pistol or virtually any centerfire caliber with a very small, light handgun, such as many that are commonly used for concealed carry.

Even with everything else being equal, when shifting to live-fire, split times will increase significantly, accuracy will degrade significantly, and the weapon itself may not even function if it is "limp-wristed." However, the same, wholly inadequate shooting mechanics may work just fine when firing simulated munitions (or laser pistols with no recoil).

If you want to see an example, try to find a functioning local arcade, maybe at a movie theater or at the mall, and look for a "first person shooter" game where somebody who is really good at the game is playing. Chances are they are absolutely crushing the game, yet using body mechanics and shooting techniques that obviously would catastrophically fail if they were firing live rounds.

The bottom line is that it is much easier to hold the weapon relatively steady and fire relatively quickly with some measure of accuracy using simulated weapons than it is with real weapons. There is quite simply much less physical force involved that the shooter needs to control.

What all of this means with respect to force-on-force training is that during rapid-fire, it is relatively easy to create a visual "stream" of rounds in the air that the visual system, mind, and body are predisposed to automatically deliver to the intended target. This will happen automatically as the brain makes corresponding performance corrections based on both the stream's trajectory and visual confirmation of the impacts.

In the real world, when shooting real bullets, not only will the mechanical and physiological requirements to make the weapon operate effectively be far more stringent (i.e., your grip and body position must be done well), but also *none* of these "fake" visual aiming systems are likely to exist. Despite the irreplaceable value of force-on-force training, its predictive ability with respect to actual *shooting* skill performance can be extremely limited, especially when the student is using non-visually-aimed-shooting techniques.

Staying purely in the realm of shooting mechanics and aiming systems, there are also at least two more significant differences between the real world and force-on-force training. First, it is very rare, especially in civilian training, for force-on-force to involve significant physical violence being done to the trainee. This is particularly true with respect to impact trauma to the head. Because of this, it is highly unusual for the trainee to perform shooting skills with a severely impaired vestibular system (the body system that runs awareness of itself and its own positioning).

The real world, however, is not so comprised. It is certainly not so during incidents requiring the application of deadly force (which typically requires the presence of imminent danger of death or serious bodily harm). Any technique that requires the vestibular system to work well will probably work a lot less well after you get punched in the head, shot, stabbed, kicked down a flight of stairs, hit with a car, etc.

Finally, in force-on-force, neither shot placement nor terminal ballistics usually matter very much, if at all. It is not impossible to implement shot placement as a factor (though it is uncommon in my personal experience, and it is very difficult to do in practice), and it is certainly not unusual, especially with marking cartridges, to evaluate hits and shot placement during a debrief. However, in most cases, a role-player will start cycling down when they are engaged successfully, regardless of the exact round placement.

The real world, however, does not work this way. Hollywood notwithstanding, real bullets, especially from handguns, do not always work that well. Actually, often they don't. Shooting, or sometimes even just shooting at, a role player—thereby causing his or her preprogrammed downcycle—is simply not the same thing as shooting a real person until they no longer pose a threat. Force-on-force is one of the best training tools we have to prepare for the real thing. That does not make it a literal equivalent.

A standard of accuracy that works every time in force-on-force training also may not work on the street. The real target area that is required to stop a real threat may be (actually probably is) a *lot* smaller than what is required for success in most simulated engagements.

We just covered a lot, but the summary is simple. In force-on-force training there is:

1. a significantly reduced standard of accuracy required for success,
2. a significantly reduced skill requirement for successful application of shooting mechanics,
3. at least two "fake" visual aiming systems that begin working at a precognitive level when eye focus and attentional focus are on the target, and
4. virtually zero chance of a compromised vestibular system that could severely impact the body's ability to conduct kinesthetic alignment.

In the real world, reverse *all* of that.

To avoid confusion, let me reiterate that non-visually-aimed-fire is an important, combat-proven skill that must be part of any functional armed skillset. Force-on-force training also is a fantastic, indispensable tool. If you are not using it appropriately and when appropriate, you are behind the curve and setting yourself and your students up for failure.

That said, if you are judging the effectiveness of *shooting* skills based primarily on performance in simulated tactical environments, especially when using non-visually-aimed-fire techniques, allow me to suggest that you are not doing yourself or your students any favors.

Chapter 8:
Training Period Performance

The next discussion point I want to address is relative end-of-training-period performance. This refers to how well students perform skills at the end of a class. Often this judgment is made based on a person's performance improvement over the course of a single training day. One of the often-used justifications for teaching non-visually-aimed-fire to entry-level shooters is that it produces much better performance in close range, practical-type shooting skill performance than visually-aimed-fire shooting does at the end of the training period. This is absolutely true in some cases. With new shooters this will certainly be true when the training structure is a one-day class. However, there is much more to the story.

Though the learning research related to this topic is outside the focus of this book, how a student performs at the end of a training period—when using training methods and structures that neurologically *cannot* produce long-term learning—is at best minimally relevant to the real world. In fact, there is significant research that suggests lower performers in these types of end-of-training skills tests will often perform *better* once they are out in the real world. To learn more about this research, I highly recommend Joan Vicker's book, *Perception, Cognition and Decision Training: The Quiet Eye in Action.*

Understand that when you teach non-visually-aimed fire as the primary, dominant skill, you are skipping a key factor of performance for aiming—one of the most important aspects of firearms use. The visual skills required for even moderate shooting performance that are neglected during neurological circuit development when using this type of training methodology with non-visually-aimed-fire also happen to be the least instinctive (or, more accurately, the least practiced on a daily basis during normal human activity) skills involved in firearms use. In other words, you are doing training that altogether ignores the single fundamental component of combat shooting that can only be adequately developed with purposeful training.

The going theory behind using this approach is that if you teach visually-aimed-fire, even with combatively applicable techniques, students will not leave their first training period with even a minimally functional skillset. This may often be true—it is almost certainly true when teaching traditional marksmanship skills first. It is simply not possible to learn combatively relevant visuomotor skills and the attentional focus required (or anything else that requires development of a new brain network) in a single day or training session. As we saw in Chapter 2, that is not how the brain works.

As discussed in Chapter 7, the body does not need to learn (or perhaps relearn) kinesthetic alignment during a training session. If the student's vestibular system is intact and functional, the alignment is something he or she is already adequately trained to do. Most students can stand within a few feet of a human-sized target (statistically within the average range of most documented gunfights) and consistently hit with "combat effective" accuracy (placing most rounds within the area of the silhouette target that corresponds to the area of the human body where gunfire is likely to be effective at stopping a threat) if they are taught to apply even a semi-functional grip while aligning their body position and shooting posture with the target.

Because students in these training environments are consistently applying assumed "combat accuracy" at "combative ranges" and can do so in relatively compressed timeframes, the assumption is that they have now developed a minimally functional defensive skillset with firearms. They are therefore considered "better off" than they were prior to the training. As a result, the training is deemed successful. Some instructors who have taught and/or seen multiple techniques and methods used in instruction also note that students trained in this manner are applying "combat accuracy" at "combative ranges" more effectively than students who would have spent that same time-period and resources learning visually-aimed-fire techniques.

There is much that is true here, and this can be a compelling set of arguments for both instructors and students. Therefore, they are worth discussing. While I will not rehash the previous chapter, it is important to point out that, even when shooting live rounds, there are still a number of significant discrepancies in square range training that impact its usefulness for evaluation of real-life shooting method effectiveness. These include the following:

1. There is a "fake" visual aiming system (visible round impacts on the target).

2. There is an artificial level of scene layout awareness (targets rarely move, or when they do move, they do so in restricted and highly predictable patterns) which contributes to an unrealistic level of kinesthetic alignment ability.

3. There is no possibility of the vestibular system being severely compromised as a prelude to the requirement for shooting skill

application (i.e., nobody is going to crack a firearms student in the head on a range).

While the differences are not as numerous or as severe here as they are with simulated munitions, they are still significant enough that performance in this setting is still not necessarily predictive of real-world shooting performance capability, even with everything else being equal.

With respect to range performance, especially for, but not limited to, beginners, it is also important to note that most of the drills and exercises during this type of training are done starting with the gun in the hand. You will probably notice, if you peruse the literature on non-visually-aimed fire, that a preponderance of the exercises conducted either start with a pre-existing grip on the gun or do not involve a measured and compressed timeframe for accessing and presenting the weapon.

There are several completely valid reasons for this. Range safety is chief among them. Rapidly drawing from the types of holsters that are actually used in the real world for most applications is frankly too dangerous to do live-fire with most of the students who are taking the types of classes where non-visually-aimed-fire is taught as the primary shooting technique. Unfortunately, this reality also *significantly* changes the relevancy of the range performance in these courses to the situations that most people are likely to encounter in the real world.

It is certainly not impossible for a person to be gifted the time necessary to administratively access and prepare a weapon, settling it into the perfect grip prior to encountering a lethal threat. For example, in a home invasion scenario, one can easily envision how this could occur. However, while this may be generally possible, it is certainly not assured. For anyone carrying a weapon either professionally or for self-defense, this situation is far less likely. Therefore, yet another a key component of non-visually-aimed-fire's close-range shooting range success becomes an unreliable variable once a person ventures off the range and out into the real world: specifically, alignment of the handgun with the shooter's arm.

When you point a finger, you are pointing part of your body. Your vestibular system, when functioning properly, can give you a clue about where you train your finger as you extend it and point it at a target. There are, however, no such systems within the inert hunk of plastic and steel that we call a firearm. The relationship of the end of the muzzle with the "natural" pointing of your arm is far from assured. Even if the arm and hand are aligned perfectly with the threat, that is only a part of the battle.

In both the real world and in high-performance, high-speed shooting, a significantly less than perfect grip is frankly not terribly unusual. This is particularly true when the pistol is accessed in haste and from a less than ideal position, such as an inside-the-waistband holster underneath both a cover shirt and a jacket, or while moving.

When clawing a gun out from beneath multiple layers of apparel under stress (once again I refer you to the video of the church shooting from Chapter 3), what are the chances that an absolutely perfect grip is assumed? What if you are running or diving for cover at the same time that you draw? How about if you are tumbling down a flight of stairs? What about getting hit in the face with a tire iron, stabbed repeatedly with a screwdriver, or charged by a person twice your size? What if your hands are slick with blood? Is the alignment of the gun with the hand still a constant?

The bottom line is that a shooter's grip on a handgun during a real fight can sometimes be less than perfectly performed, and the alignment of the barrel axis with the arm can vary significantly while still permitting the shooter to operate the weapon. Whether or not the grip is performed well, though becomes somewhat academic in a real fight. The grip that you perform simply is—you still may only have the rest of your life to solve the problem.

You can get an idea how a less than ideal (yet still functional) grip can impact the results of non-visually-aimed-fire right now if you have access to a pencil, pen, straw, or similar object. Even a cell phone would work. Take one of your hands and hold it in front if you as if you were going to perform a handshake. Now drop your thumb so that it is parallel with the floor, forming a "V" with the webbing of your hand.

With the other hand, place the back end of your chosen object, I will assume it is a pencil, directly in the middle of the "V" and point your hand (and thereby the front of the pencil) at a chair (or other similar sized object) about four feet away from you. It should be easy to see how you can point your arm and fingers at the chair and easily hit it if a projectile were to be fired from the pen or pencil.

Now shift the back of the pencil either left or right without moving your "shooting" arm or hand. How far do you need to move it before the front of the pencil is no longer pointed at the center of the chair? How far before it is no longer pointed at the chair at all? Move back a few paces and repeat this exercise. How far away do you need to be before even a minor irregularity in grip alignment renders the body's kinesthetic pointing ability ineffective for combative shooting purposes?

I will say it one more time: Non-visually-aimed-fire is an important skill with important applications and has been used successfully to win gunfights for at least a century. This does not mean, however, that the results of performing the technique during static range drills against paper targets (especially if a shooting grip is assumed prior to the drill or without haste, impedance, and distraction) is in any way representative of how the technique will work in combat.

It is also important to point out that, much like the failures associated with much of police training, end-of-training-period assessments are based on evaluating the results of a fundamentally flawed training methodology that cannot work at a neurological level. Most common training structures are only capable of making students adapt that which they already know. The training itself cannot make them learn.

Looking purely from a training method and technique application standpoint (ignore the shooting subject matter for a moment), the pitch for this sort of training methodology could be summarized as follows:

We exclusively use a training structure that does not work well because it fails to match how the human brain learns. Because of our flawed training structure, we cannot teach you to do things that you do not already know. We assume that you are probably never going to practice. We also assume that you do not want to, will never want to, and do not need to be knowledgeable of the subject matter you are asking us to teach you.

Therefore, instead of teaching you things that work well, we are going to show you how to adapt things you already know to get the right answers some of the time. What we will teach you will not solve some of the problems you are here to learn how to solve. It will also prevent you from learning the other techniques that will help you solve them.

After our class, you probably will not be able to learn the things that will solve the problem you came here to learn to solve, even if you practice on your own a lot and spend a lot of money.

In fact, the more you practice what we teach you, the harder it will be for you to learn the skills that work much better and will solve your problems.

Do not worry, though! We know better than you and think that you are not special enough to need to be able to solve the problems that you came here to learn how to solve. We are the experts, after all. We know better. You do not know anything, and that is why you are here.

Sign up today!

Any volunteers? Yet this is exactly what teaching non-visually-aimed-fire to new shooters and having them repeatedly hose down targets without integrating visual skills at close range does. If you do this with your students, or if you attend a training course where this is done, that narrative accurately describes the training experience.

I recently received an email (in Spring 2021) from one of the first law enforcement academies in the U.S. that implemented brain-based training design. They started nearly ten years ago, and their results showed exponential improvement, dropping their failure rates from near 12% to consistently less than 1%, even as their program became far more difficult. They have continued to refine the methodology in implementation and are constantly refining their methods as they continue to learn from student outcomes. Here is an excerpt from the email:

> Unblocking training, firearms, defensive tactics, first responder, etc., has been one of the best things we've done for law enforcement training and our academy.
>
> We just finished an academy class with 100% passing in firearms. *Those that struggle most have prior so-called "training" which impedes their development* [emphasis added].

Does any student want to attend firearms training, often at considerable personal expense, only to have that the results of that time and effort *become an impediment* not only to developing a successful career as an armed professional, but also to his or her personal capability, potential to develop as a shooter, and possibly even personal survival on the street? Does any instructor want to provide the training that makes it nearly impossible for a student to succeed, either professionally or, even worse, in armed combat?

I have a former student who is now in federal law enforcement. This student contacted me—stunned—during academy firearms training. The student was shocked not only by the non-challenging and tactically irrelevant performance standards and ineffective training methodology being applied (at great taxpayer expense), but also that several of the other students who were struggling to even pass the firearms program had dedicated significant time and resources to developing shooting skill prior to attending. One classmate in this situation had apparently spent over $60,000 on firearms training before the academy! What was the result of this time, money, dedication, and effort? A poor, ineffective skillset that actively *prevented* him or her from learning anything else.

One of the great limitations that exists in the firearms and tactical training industry today is that very few instructors have long-term equity in the performance of their students. In the vast majority of cases, instructors simply see students for a brief, blocked training period, then not for months or years afterwards, if ever again.

Many instructors also focus on teaching, running, and evaluating a specific set of skills or specific drills and therefore naturally end up with training methods that optimize some combination of end-of-training period performance and, especially for civilians, the "fun factor" of the training. There is nothing inherently wrong with either of these objectives. Unfortunately, however, this often does little to positively impact long-term student learning or the development of anything the student does not already know how to do before the class starts.

To the instructors: Allow me to suggest that, while there are legitimate reasons to train this way with non-visually-aimed-fire (such as the reasons this training was developed by Applegate, Sykes, and Fairbairn), at the very least, you should be disclosing to your students the impediments to future skill development and the limitations in operational capability that may be imposed by the structure of your training methods.

Some students might be OK with it. Some may even prefer it, which is perfectly acceptable. However, I believe that every student should absolutely be given the opportunity to provide fully informed consent before an instructor applies training methods that negatively, and perhaps permanently, limit his or her potential to perform. We must acknowledge the limitations of what and how we teach—and everything has limitations. Our students should also be provided with a choice, as should we.

If you must introduce or learn shooting this way for legitimate tactical reasons, consider clearly defining the limitations of the shooting method as well as the reasons for it being taught as it is to the students. The brain, when self-aware, can be quite powerful in its ability to regulate its own development. In this case, at the earliest opportunity, I strongly suggest reteaching shooting skills in such a way as to make combatively valid techniques for visually-aimed-fire the dominant skill response.

Chapter 9:
Market Forces

One of the last chapters in the second volume of my *On Training* series outlines what I believe are the systemic student motivations for seeking out and attending training in the civilian market. The elevator summary is that most people only go to training because they are required to by law for a carry permit. Of those students who do attend training programs of their own volition, the most prevalent personal motivations are entertainment and getting a cool "bragging rights" experience. The least common motivations are education and skillset development.

Looking beyond even the training industry, ask yourself what the average person values the most. Without getting too philosophical, in at least my own personal observations and experience, the most important human motivator is usually a person's desire to feel good about himself or herself.

People like to succeed. They hate to fail. They especially hate to fail when they are spending their money and, often more importantly, their leisure time on the effort. While this is greatly oversimplified, people like receiving hits of dopamine and oxytocin (brain hormones produced by pleasurable, rewarding experiences). We therefore seek out opportunities to release these chemical compounds into our brain tissue.

Unfortunately, highly effective processes for developing robust and functional combative skillsets do not typically start out with a massive release of pleasure chemicals. Many well-intentioned, but poorly designed, programs even start out by inducing significant amounts of stress chemicals instead. This turns many students off and, while this is a topic that is outside the scope of this book, it also makes their ability to learn far less effective.

Even well-designed programs, especially at the beginning, will not necessarily be a barrel of fun. Effectively learning nuanced motor skills with complex brain maps (such as shooting and fighting) requires repetition over time. There is just no way around it. It is how the brain works, fundamentally.

Distributed learning over time is the best way to learn. It is also the fastest way to functional skills. These can then become a lot of fun to practice and apply in training, as well as functional in the real world. Though this approach works effectively, it is not, at least at the beginning, the most sexy, high-speed, fun, or "cool experience."

Who has the most competitive market advantage? Is it an instructor who provides students "success" without effort or learning? Congratulations! You just emptied a magazine in 3 seconds at 4 feet into a static piece of paper while using a typically undisclosed "fake" visual aiming system! Go home and feel good about yourself!

Or is it an instructor who provides students with a structured approach to skill development that matches how their brains learn and develops robust and functional skillsets? As an instructor, how do you help keep your students stay alive or develop professionally functional skills when they either really just want to have some fun and blow off some ammo or, alternatively, "just want to defend myself" and have no interest in actually acquiring the minimal skills necessary to do so?

It is an interesting question. Unfortunately, it is one that many people in the industry answer with a false choice. Many people, including far too many instructors, assume that in order to achieve functional defensive firearms skills one must first become a shooting enthusiast. This simply is not true.

Being enthusiastic about shooting will certainly help provide motivation and enjoyment; however, it is not a prerequisite. A person does not need to enjoy firearms to learn to use them effectively. Someone just needs to be willing to put forth a little bit of time and effort aligned with how the brain learns skills and information.

Unfortunately, there are very few places in the civilian world where you can currently receive firearms training in a way that matches how the brain learns. While brain-based training works extremely well, in most cases the business model fails. It is my hope that, as more options become available in the market (such as martial arts models of training, and the integration of hybrid learning programs that use effective, relevant technology to match information delivery to how the brain learns) and consumers begin to see and understand that they have a choice (and understand the choice), the structure of the training industry will change for the better.

Some people will always just want to blow off ammo—during times when it is affordable and available—and have fun, which is great. However, there are enough people who do take skill-at-arms more seriously and consider it, even if they enjoy the challenge, something other than purely a recreational activity. This does not require enthusiasm for the sport. It only requires recognition that the capability for individual physical equality is thing worthy of possession.

I believe that many people would take it seriously if there was a real, viable choice in training. I also believe that, as in most industries, the market is capable of supporting a plethora of options. As long-term comparisons of results start becoming clear (you cannot hide on a range), the choices will become starker as well.

Chapter 10:
A Matter of Training

I previously mentioned that a reader once asked me whether I had studied the literature proving that visually-aimed-fire was impossible in combat. Consider that the average adult in the United States could not do a single pull-up if asked to perform one on demand. Yet performing a pull-up is clearly not impossible. Doing a single pull-up is not even difficult, given a minimal amount of appropriate preparation and training.

What might a human performance study across the aggregate population of the United States determine about pull-ups? It is likely that it would identify pull-ups as "impossible." The humble pull-up could well find itself classified as a skill so rarely performed successfully that it falls outside the normal parameters of human genetic capacity.

This is obviously false, but only because we experientially know better and because pull-ups are an activity that we can see performed visually in front of us, unlike another person's visual skill performance. A better designed study might incorporate evaluation of each person's habits and exercise routine. This study might determine that virtually every healthy human being possesses the capability of doing pull-ups. The study might also determine that a fairly minimal exercise routine can bring to the average person the ability to do a pull-up in a very reasonable timeframe.

We should take a similar viewpoint when it comes to considering the application of visual skills in combat shooting. When evaluating the development of non-visually-aimed-fire as a shooting technique, it is worth noting that Applegate, Fairbairn, and Sykes only had a fraction of the resources (time being by far the most valuable) that are available now, even to civilians and armed professionals in the most austere of training environments. Is anyone today dropping students behind enemy lines within a week of training?

Skill performance is driven from the *human* system, not from specific components of hardware, such as sights on a gun. Without human performance, the hardware is largely irrelevant. If students, unknowingly, develop a primary shooting technique that has a poor chance of solving some important tactical problems that are relevant in today's world because of how an instructor taught them, I consider this to be a serious ethical issue for the training industry.

It is certainly possible to hit a target without the use of a visual sighting system. It depends on the target and the conditions. There comes a point, however, depending on a host of variables such as what the target is, how far away the target is, the state of the shooter's vestibular system, the shooter's kinesthetic development from the specific shooting position being used, relative motion of the target, etc. where that ability stops. The use of a visual sighting technique (along with adequate trigger manipulation) then becomes necessary to solve the tactical problem.

Once a person has learned to shoot at a procedural level, if the necessary visual skills do not "automatically" happen, chances are that they will not ever be performed outside of a low stress training environment. In many cases these skills will also be effectively unlearnable for combat applications once something else has been developed as the dominant shooting technique. *This* is the trouble.

Is it possible to aim in a gunfight? Or is it impossible? The truth, based not only on years of teaching, working with, and seeing shooters of every caliber and pedigree perform in a wide variety of settings but also based on years of in-depth research into what is currently known about how the human brain works, is that *both* statements are correct. The determining factor is how a student received their initial training and exposure to shooting.

The challenge from a training design perspective is not teaching various shooting techniques, nor is it understanding which techniques are appropriate for solving specific tactical problems. (Note that there may often be more than one acceptable answer here.) Rather, the challenge is developing the student's skillset so that the "right" combination of techniques for solving the tactical problem are the techniques that get applied automatically in the real world.

This brings me right back to our flagship issue at Building Shooters, which is addressing the first of the two big failures in today's training industry. This failure is not based primarily in teaching flawed technique, nor, necessarily, in flawed tactical application of various techniques. It is based in flawed *training design and method*. It is our industry standard method of training delivery that is broken.

When new shooters are introduced to shooting, and repetitively practice shooting, without learning—or often frankly even learning about—the necessary visuomotor skills and attentional focus for applying visually-aimed-fire in combative settings, both their immediate personal skillset and their long-term performance potential are severely compromised.

Students do not really learn shooting skills this way. Like the basic-level students of Applegate, Fairbairn, and Sykes, these students are taking something they already know (how to align their body in a targeting task using gross motor skills) and applying it to perform a skill that can kill effectively with a firearm at very close ranges, but that falls apart when more is needed.

The negative long-term performance impacts of teaching entry-level students non-visually-aimed-fire and having them practice it almost exclusively are severe. These effects can become virtually irreversible, which I believe represents an ethical issue in the training industry when this is not clearly defined and disclosed to the student ahead of time. This training delivery failure does not just build shooters who cannot aim in gunfights—it builds shooters who *cannot learn* to aim in gunfights.

I acknowledge some hyperbole here. Is it possible to overcome the negative effects of this training structure and attain an effective combative skillset after the fact? Yes, in most cases I believe it is. But it will be very difficult and will require a far more significant investment of both time and resources than simply developing an adequate skillset to begin with.

It is simply a cold, hard fact that a person who trains to apply non-visually-aimed-fire as his or her primary mechanism for hitting the target has embraced (whether knowingly or not) a significant distance and accuracy limitation that severely limits his or her capacity to successfully employ a firearm. It is equally true that the majority of students will never recover from being initially trained with this type of methodology, even if they spend a lot of time training. Some of them can with (literally) therapy-level interventions. But most of them will not.

If you teach this way and your students actually practice what you teach them, it becomes more difficult for them to acquire a functional and robust combative skillset every time they train. I am not saying that you should not teach non-visually-aimed-fire. You should teach it. I *am* saying that it should *not* be developed as the dominant shooting technique without the informed consent of the student.

Instructors who teach this way (again—there are legitimate reasons to do so) should start disclosing to students what they are doing, why they are doing it, and what the negative impacts to the student are likely to be. In my opinion, each student should also provide informed consent, in writing, before being trained using methods that will both hurt their performance potential as shooters and possibly limit their ability to defend themselves and their families.

Recommendations and considerations for training

Criticism is great; however, it is of limited value if there is not some accompanying, constructive action. Therefore, based on what I know right now, here are my current suggestions.

First, if you are building a skillset intended for combative applications—or that may ever need to integrate with combative applications—it is *imperative* that you develop a student's combat-relevant visual skills as a structural component of their primary shooting skill. If you do not, you set them up for severe limitations in their potential to perform. This could impact even their ability to effectively defend themselves and their family, and it will certainly have a negative impact on any attempt at a career as an armed professional.

Do not let anyone tell you that it is impossible for a person to use the sights in a gunfight or under stress. It requires training, but it is more than possible. In fact, it may even require less training and practice to learn than pointing does—we just happen spend a lot of time training to point as we move through and interact with the world on a day-to-day basis. As discussed in the previous section, however, it is likely that once non-visually-aimed-fire is consolidated as the dominant shooting skill, both using the sights in a gunfight and *learning* to use the sights in a gunfight will become, more or less, impossible.

Second, do not introduce non-visually-aimed-shooting techniques with any level of repetition until *after* the visually-aimed-skills are wired into the student's brain (learned, procedurally consolidated, and integrated with the "machine"). If you do, you are introducing a "substandard" competing skill with what is almost certainly a lower energy performance requirement. This will make it very difficult to predict and control which technique becomes dominant.

Third, consider the reality of the "fake" aiming systems that are integrated into most of our common training methods, with range training, simulators, and force-on-force. With the increasing popularity of laser-based, dry-fire tools, this issue now even extends into the dry-fire realm.

There is nothing wrong with any of these tools. They are great tools. I use them. Most shooters who are very good (including the many who are far better than I am) use them. These tools simply have limitations. They also have the potential to develop dependency for performance on a visual aiming system that does not exist in the real world. When training with these tools, guard yourself and your students against the temptation to slide into applying "easier" visual techniques that depend on stimuli that will not exist for real.

One way to do this is to demand standards of accuracy that preclude application of the "fake" visual aiming systems. Your favorite shooting drills? Instead of performing them at 3-7 yards, do them at 15 yards using an 8-inch pie plate or sheet of notebook paper as the target. Get hits. Do not accept missing just because of the distance.

Instead of shooting at 3- or 6-inch dots, consider shooting at 3- or 6-inch holes—only misses provide a fake visual stimulus. This is especially effective indoors with a plate trap where you cannot see impacts on a berm. Another way to avoid these outcomes, of course, is to apply extreme mental discipline during training.

Consider that preparing for the fight is not about replicating the distance and round count of the "standard" engagement. Who cares about that? Is anyone likely to ever be in imminent danger of death or harm from a static piece of close-range cardboard?

Preparing for the fight is about preparing the same neurological and physiological machinery that is required for success in the real world. Applying "fake" visual aiming systems at ranges matching the aggregate gunfight distance does not fit this bill. Developing the use of a *real* visual aiming system as the dominant, procedural response does.

Fourth, develop non-visually-aimed-shooting skills as responses that are associated with specific relevant stimuli and context. I will not discuss specific techniques or tactical applications here. If you are an instructor who can and should be teaching this, you should already know what this refers to.

Please note that, as of the publication of this book, Building Shooters (the company) is several months away from publicly launching our revolutionary NURO™ Shooting System (Patent Pending). This elegantly simple, yet extraordinarily powerful set of training tools will forever alter the landscape of firearms and tactical training. NURO™ will change what is possible to achieve, particularly with respect to decision-making and visual skills. Please visit our website at www.buildingshooters.com and sign up for our newsletter to stay up-to-date on product availability.

Fifth, consider developing non-visually-aimed-shooting skills so that they are only associated with motor skill performance that differs from that of the dominant shooting response. In other words, applying the "bread and butter" shooting motor skills, shooting positions, etc., *never* (an overly strong word) involves shooting without the visual reference(s) except in exigent situations (sights broken off, etc.).

At least in theory, non-visually-aimed-shooting then becomes a unique brain map, associated with its own unique set of tactically relevant stimuli. This should greatly reduce the chances that non-visually-aimed-fire is performed as a dominant response to a real-world stimulus that can only be addressed effectively with the appropriate application of visual skills as part of the shooting cycle.

Sixth, and finally, work to educate yourself, your students, and others in this industry about the issues that exist both with training and technique limitations. The human brain is incredibly powerful. It possesses the ability to change itself in remarkable ways when required. This is especially true once it becomes "self-aware."

If you are "buying what I am selling" here (not books or products, but training concepts), please do not keep it to yourself. Join us at Building Shooters in our efforts to remake this industry, especially the combative and tactical side.

Help us transform this profession into a discipline marked by efficiency, effectiveness, and continuous improvement, not entertainment and instant gratification. Help people to understand their own motivations and the consequences of the choices they make about how they choose to spend their training time and training money.

Finally, give your students (and yourself) the opportunity to make an informed decision about skillset and survivability.

Break through the limits of dogma that is rooted in the limitations and assumptions of the past. When training for the real world, pursue your best chance of Hitting in Combat.

Thank You and Join Us!

Thank you for reading this book. We hope you found it enlightening and have taken with you some ideas that can help improve immediately both the quality and results of your training programs.

If you found this book valuable, please take a moment and post a short review of it on Amazon to help us spread the word about the value of brain-based training.

If you have not yet done so, please also take this opportunity to sign up for access to our newsletter and our free comprehensive video series on gun safety and gun handling fundamentals. Join us in working to make the industry, and society as a whole, a better, safer place. Access is available at:

https://www.buildingshooters.com/free

References and Recommended Reading

Aivar, M. P., Brenner, E., & Smeets, J. B. (2015). Hitting a target is fundamentally different from avoiding obstacles. *Vision Res, 110*(Pt B), 166-178. doi: 10.1016/j.visres.2014.10.009

Applegate, R. (1976). *Kill or get killed: Riot control techniques, manhandling, and close combat, for police and the military* (New rev. and enl. ed.). Boulder, CO.: Paladin Press.

Applegate, R., & Janich, M. D. (1998). *Bullseyes don't shoot back: The complete textbook of point shooting for close quarters combat.* Boulder, CO.: Paladin Press.

Archambault, P. S., Ferrari-Toniolo, S., Caminiti, R., & Battaglia-Mayer, A. (2015). Visually-guided correction of hand reaching movements: The neurophysiological bases in the cerebral cortex. *Vision Res, 110*(Pt B), 244-256. doi: 10.1016/j.visres.2014.09.009

Bergland, C. (2014). Is Impaired Neuroplasticity Linked to Chronic Pain? New research links chronic pain with impaired neuroplasticity of the brain. Retrieved from https://www.psychologytoday.com/us/blog/the-athletes-way/201403/is-impaired-neuroplasticity-linked-chronic-pain

Biederman, I., Mezzanotte, R. J., & Rabinowitz, J. C. (1982). Scene perception: Detecting and judging objects undergoing relational violations. *Cogn Psychol, 14*(2), 143-177. doi: 10.1016/0010-0285(82)90007-x

Bukalo, O., Campanac, E., Hoffman, D. A., & Fields, R. D. (2013). Synaptic plasticity by antidromic firing during hippocampal network oscillations. *Proc Natl Acad Sci U S A, 110*(13), 5175-5180. doi:10.1073/pnas.1210735110

Burr, D., & Thompson, P. (2011). Motion psychophysics: 1985-2010. *Vision Res, 51*(13), 1431-1456. doi: 10.1016/j.visres.2011.02.008

Calvert, G., Spence, C., & Stein, B. (2004). *Handbook of multisensory processing*. Boston: MIT Press.

Cavanagh, P. (2011). Visual cognition. *Vision Res, 51*(13), 1538-1551. doi: 10.1016/j.visres.2011.01.015

Cluff, T., Crevecoeur, F., & Scott, S. H. (2015). A perspective on multisensory integration and rapid perturbation responses. *Vision Res, 110*(Pt B), 215-222. doi: 10.1016/j.visres.2014.06.011

Davenport, J. L., & Potter, M. C. (2004). Scene consistency in object and background perception. *Psychol Sci, 15*(8), 559-564. doi: 10.1111/j.0956-7976.2004.00719.x

Doidge, N. (2007). *The brain that changes itself: Stories of personal triumph from the frontiers of brain science*. New York: Viking.

Doidge, N. (2016). *The brain's way of healing: Remarkable discoveries and recoveries from the frontiers of neuroplasticity* (Updated and expanded edition. ed.). New York, New York: Penguin Books.

Enos, B. (1990). *Practical shooting:Bbeyond fundamentals*. Clifton, CO: Zediker.

Fairbairn, W. E., & Sykes, E. A. (1974). *Shooting to live, with the one-hand gun*. Boulder, CO: Paladin Press.

Fields, R. D., & Ni, Y. (2010). Nonsynaptic communication through ATP release from volume-activated anion channels in axons. *Sci Signal, 3*(142), ra73. doi:10.1126/scisignal.2001128

Fields, R. D. (2014). Neuroscience. Myelin—more than insulation. *Science, 344*(6181), 264-266. doi: 10.1126/science.1253851

Fields, R. D. (2020). The Brain Learns in Unexpected Ways: White matter, the insulation around our neural wiring, plays a critical role in acquiring knowledge. *Scientific American.* Retrieved from https://www.scientificamerican.com/article/the-brain-learns-in-unexpected-ways/

Fize, D., Cauchoix, M., & Fabre-Thorpe, M. (2011). Humans and monkeys share visual representations. *Proc Natl Acad Sci U S A, 108*(18), 7635-7640. doi: 10.1073/pnas.1016213108

Gilmartin, B. (1986). A review of the role of sympathetic innervation of the ciliary muscle in ocular accommodation. *Ophthalmic Physiol Opt, 6*(1), 23-37. Retrieved from https://www.ncbi.nlm.nih.gov/pubmed/2872644

Grill-Spector, K. (2003). The neural basis of object perception. *Curr Opin Neurobiol, 13*(2), 159-166. doi: 10.1016/s0959-4388(03)00040-0

Haydon, P. G., & Carmignoto, G. (2006). Astrocyte control of synaptic transmission and neurovascular coupling. *Physiol Rev, 86*(3), 1009-1031. doi:10.1152/physrev.00049.2005

Hughes, E. G., Orthmann-Murphy, J. L., Langseth, A. J., & Bergles, D. E. (2018). Myelin remodeling through experience-dependent oligodendrogenesis in the adult somatosensory cortex. *Nature Neuroscience, 21*(5), 696-706. doi: 10.1038/s41593-018-0121-5

Krigolson, O. E., Cheng, D., & Binsted, G. (2015). The role of visual processing in motor learning and control: Insights from electroencephalography. *Vision Res, 110*(Pt B), 277-285. doi: 10.1016/j.visres.2014.12.024

Kuang, S., & Gail, A. (2015). When adaptive control fails: Slow recovery of reduced rapid online control during reaching under reversed vision. *Vision Res, 110*(Pt B), 155-165. doi: 10.1016/j.visres.2014.08.021

Lamb, K. E. (2011). *Stay in the fight! Warrior's guide to the combat pistol*. Trample & Hurdle.

Lourenço, J., De Stasi, A. M., Deleuze, C., Bigot, M., Pazienti, A., Aguirre, A., . . . Bacci, A. (2020). Modulation of Coordinated Activity across Cortical Layers by Plasticity of Inhibitory Synapses. *Cell Reports, 30*(3), 630-641.e635. doi: https://doi.org/10.1016/j.celrep.2019.12.052

Lu, Z. L., & Sperling, G. (1995). The functional architecture of human visual motion perception. *Vision Res, 35*(19), 2697-2722. doi: 10.1016/0042-6989(95)00025-u

Mackrous, I., & Proteau, L. (2015). Is visual-based, online control of manual-aiming movements disturbed when adapting to new movement dynamics? *Vision Res, 110*(Pt B), 223-232. doi: 10.1016/j.visres.2014.05.007

Mackrous, I., & Simoneau, M. (2014). Generalization of vestibular learning to earth-fixed targets is possible but limited when the polarity of afferent vestibular information is changed. *Neuroscience, 260*, 12-22. doi: 10.1016/j.neuroscience.2013.12.002

McNamara, P. (2008). *Tactical application of practical shooting* (2nd ed). iUniverse

Munneke, J., Brentari, V., & Peelen, M. V. (2013). The influence of scene context on object recognition is independent of attentional focus. *Front Psychol, 4*, 552. doi: 10.3389/fpsyg.2013.00552

Murray, K. R. (2004). *Training at the speed of life, volume one: The definitive textbook for military and law enforcement reality based training* (1st ed.). Gotha, FL: Armiger Publications.

Pincus, R. (2010). *Combat focus shooting: Evolution 2010*. Virginia Beach, Virginia: I.C.E. Publishing Company.

Quail, J., Wollert, T., (2018). *A scientific approach to reality based training*. Setcan Corp.

Salomon, D. P. (2016). *Building shooters: Applying neuroscience research to tactical training system design and training delivery*. Silver Point, TN: Innovative Services and Solutions LLC.

Salomon, D. P. (2016). *On training: Selected essays*. Silver Point, TN: Innovative Services and Solutions LLC.

Salomon, D. P. (2016). *On training, volume 2: Selected essays*. Silver Point, TN: Innovative Services and Solutions LLC.

Schwab, M. E., & Strittmatter, S. M. (2014). Nogo limits neural plasticity and recovery from injury. *Curr Opin Neurobiol, 27*, 53-60. doi: 10.1016/j.conb.2014.02.011

Schwartz, J., & Begley, S. (2002). *The mind and the brain: Neuroplasticity and the power of mental force*. New York: Regan Books/HarperCollins.

Somogyi, P., Tamás, G., Lujan, R., & Buhl, E. H. (1998). Salient features of synaptic organisation in the cerebral cortex. *Brain Res Brain Res Rev, 26*(2-3), 113-135. doi:10.1016/s0165-0173(97)00061-1

Stein, B. E., & Stanford, T. R. (2008). Multisensory integration: Current issues from the perspective of the single neuron. *Nat Rev Neurosci, 9*(4), 255-266. doi: 10.1038/nrn2331

Stevens, A. P. (2014). Learning rewires the brain: In the process, some of the brain's nerve cells change shape or even fire backwards. *Science News for Students*. Retrieved from https://www.sciencenewsforstudents.org/article/learning-rewires-brain

Vickers, J. N. (2007). *Perception, cognition, and decision training: The quiet eye in action*. Champaign, IL: Human Kinetics.

Winn, B., Culhane, H. M., Gilmartin, B., & Strang, N. C. (2002). Effect of beta-adrenoceptor antagonists on autonomic control of ciliary smooth muscle. *Ophthalmic Physiol Opt, 22*(5), 359-365. doi: 10.1046/j.1475-1313.2002.00075.x

Yu, W.-M., Appler, J. M., Kim, Y.-H., Nishitani, A. M., Holt, J. R., & Goodrich, L. V. (2013). A Gata3–Mafb transcriptional network directs post-synaptic differentiation in synapses specialized for hearing. *eLife, 2,* e01341. doi:10.7554/eLife.01341

More from Building Shooters

If you would like to learn more about firearms and tactical training, applied neuroscience, and how the firearms training industry can improve, please consider the following additional titles from Building Shooters:

Building Shooters: Applying Neuroscience Research to Tactical Training Design and Training Delivery

Mentoring Shooters: The Gun Owner's Guide to Building a Firearms Culture of Safety and Personal Responsibility

On Training: Volume 1, Selected Essays

On Training: Volume 2, Selected Essays

Becoming Shooters: A Guide for New Gun Owners (Available May 2021)

NURO: A Brain-Based Analysis of Tactical Training and the Basis of Design for the World's Most Capable Tactical Training System (Available May 2021)

Building Shooters books are available at www.buildingshooters.com, on Amazon.com, and from other online book retailers.

NURO:

A Brain-Based Analysis of Tactical Training and the Basis of Design for the World's Most Capable Tactical Training System

Excerpt:

Building Shooters (the Company) is founded on the concept of using applied neuroscience and psychological research to develop brain-based training systems that improve efficiency and effectiveness in firearms and tactical training. Our intent is to facilitate improvement in performance and outcomes for the armed professions through better methods, tools, designs, and applications. The objective of this book is to define the theory and conduct a brain-based analysis of tactical training requirements that form the basis of design for Building Shooters' **NURO™ Shooting System (Patent Pending)**.

The genesis of **NURO™** occurred during a discussion regarding the pros and cons of existing training objectives, methods, and technologies. These include video-based simulators, the use of shot-timers during both live-fire and dry-fire training, the use of audible stimuli to define what targets a student should engage during a training evolution, the use of turning targets, and force-on-force training, among others. Each of these training methods have their own capabilities and limitations. Each also can produce operationally manifested "training scars" when either used improperly or when used exclusively as a training method.

During an after-action analysis of this discussion, a single and consistent issue with each of these (and other) training methods and tools became clear. *None* of the current training and qualification methods commonly used in the firearms industry today are capable of functionally engaging the same neurological and physiological mechanisms required during a real-world application of deadly force— *within the parameters necessary for effective operational performance development.*

Most of the existing tools and methods will not do the job at all. They cannot. They are purely focused on developing and measuring discreet skill and skill sequence performance in isolation. Those that can stimulate some—or even all—of the relevant functions are not scalable or cost effective enough to facilitate the volume of repetitions necessary to consistently produce effective learning and long-term retention.

The fundamentals of cognitive-infrastructure-based (connectionist) learning theory indicate that development and improvement (i.e., learning) occurs most effectively, predictably, and controllably through repetitive use of the relevant neural circuitry. Understanding this and the factors that drive it is critically important in an engineered training context. Application of these concepts facilitates a method of training design that is comparable to high-performance computing development. It is high efficiency, "bare metal" software development directly in the relevant components of the processor (brain).

Connectionist learning theory is based on the principles of neuroplasticity and considers learning at the level of basic neurological function where new connections formed between neurons account for learning. At a systems level, preparing for (learning for) a use of force encounter is, neurologically, a matter of creating an efficient brain map that corresponds to the brain map requirements of the encounter itself. The capability of efficiently and predictably creating these brain maps should therefore be considered one of the most significant factors that is relevant to a training system's ability to prepare students for successful operational outcomes.

The training tools and methods that are currently in predominant use throughout the industry fall short in this area. They are either incapable of activating the full brain map relevant to deadly force encounters, or they make it functionally impossible for any one student to perform the number of repetitions over time necessary to *develop* the applicable brain map(s) for successful critical incident performance.

Preparing students for successful operational outcomes is, therefore, extraordinarily difficult. It is also rarely predictable or consistent in terms of results, at least outside of high-attrition, high-resource training environments such as those involved in the selection and training for elite units. Outside of these highly specialized environments, armed workforce preparation and management can be summarized as using tools that do not work to prepare the wrong physiological and neurological machinery to achieve a meaningless and irrelevant standard of performance.

It should not, therefore, surprise anyone that most armed workforces such as police and security officers leave a great deal to be desired when it comes to use-of-force performance. Given the circumstances of their training, it is surprising that they achieve as much success as they do in these areas. As a factual matter, most members of conventional armed workforces who can legitimately be called competent in use-of-force skills have developed their ability almost exclusively on their own time, and at their own expense.

The **NURO™ Shooting System** was developed to provide armed workforces and others the tools necessary to address these systemic failures. **NURO™** is comprised of a set of cost-effective, scalable tools for the firearms and tactical training industry that facilitate high-repetition stimulation of, use of, development of, and enhancement of the relevant brain maps and physical skills for use of force encounters.

Rather than simply being a training tool, the **NURO™** system is the foundation for fundamentally and radically restructuring the concepts of training performance and firearms qualification. With proper application, it will facilitate operationally relevant, high-repetition training—over time—at every level of the industry, in any training setting, and within virtually any budgetary constraint...

...The foundational legal, liability, and policy basis for placing armed professionals into service, whether military, law enforcement, or security, is training and qualification. Each agency or organization has a training and qualification structure and system that it uses to certify, usually on a periodic basis, that individuals are, in fact, trained and prepared to perform the "armed" portion of their job-related duties.

Because firearms qualification is the measurable component of the foundational basis for establishing a person's ability to perform the "armed" part of the armed vocation, qualification structures are usually based on assessed "job related" skill performance requirements. These assessments of job-related skills are often designed based on the available operational performance data. For example, at what range do real-world shootings happen and how many rounds are fired at each distance?

Qualification standards also are often generally distributed based on statistical applicability of job-related use at the time the test was created.

For example, an agency may historically assess that 75% of the rounds fired by its employees in the line of duty are fired from within 3 yards of the target. As a result, that agency may mandate that about 75% of the rounds fired during qualification must be from within a distance of 3 yards, matching the historical operational performance requirements.

This approach is generally thought to reduce organizational liability and, indeed, historically agencies have suffered in employment-based litigation if difficult performance requirements cannot be established to be job-related tasks. This is because employment-related tests must typically be shown to be both job-related and scientifically valid. For example, if an agency required 75% of the rounds fired on its qualification to be from 25 yards, yet 75% of the actual historical rounds fired in the field during the agency's history were from inside 3 yards, that agency could potentially face a difficult time in litigation if an employee or recruit was terminated or faced punitive action based on failure to perform shooting tasks at 25 yards.

We agree with the general notion that both training and the associated measurement/qualification should be based on job-related-performance requirements. Unfortunately, the way in which this is attempted in virtually all existing firearms training systems is ineffective and does little to functionally address *operational* liability.

The reality is that there is little, if any, historical correlation between training performance measurement (qualification score) and operational outcomes or shooting performance in the field. This fact alone should give the industry some pause, although it does not, primarily—at least in our view—because there has never been a viable alternative approach to qualification for large agencies besides the methods that currently exist. Prior to the development of the **NURO™**

Shooting System, there has never been a method of measuring *actual* job-related-performance.

There is, however, more than simply a lack of correlation from testing performance to operational outcome to indicate that today's commonly used approaches to qualification more than miss the mark. Through conducting the basic analysis of the physiological and neurological requirements for real-world tactical skill performance, such as that contained in this book, it can easily be seen that the qualification structures predominantly in use today do not engage the same physiological and neurological mechanisms that are involved in real world use of force events.

Almost all qualifications require a physical response to a simple, pre-known stimulus in isolated settings, followed by performance of a pre-known motor skill sequence. Qualifications also predominantly involve the use of the audio system for receiving a simple stimulus. This is followed by immediate and direct access to the declarative memory system (long-term conscious access), and, finally, corresponding performance of the pre-known (and practiced) motor skill sequence (such as drawing and firing four rounds).

Compare this to the neurological and physiological performance requirements of use of force incidents. It quickly becomes obvious that, although distances and percentages of rounds fired at each distance during qualifications may match aggregate historical gunfight data, firearms qualifications do not actually involve the performance of job-related skills.

Most qualification fire is, neurologically and physiologically, mostly *unrelated* to actual operational (job-related) skill performance. The technical skills used are frequently different, the memory system from which the information is recalled is different, and the information processing centers and methods are different. It should, therefore, come as no surprise that qualification performance has little correlation to gunfight performance or other corresponding operational outcomes.

Qualifications involve predominant use of shooting and gun handling skills that are, at best, minimally applicable to the real world. The presence of non-operationally-viable visual aiming systems (such as bullet impacts on paper at close ranges and closing one eye during aimed-fire) and the normal separation of firearms skills from defensive tactics skills, especially at close range, are examples of skills and skill sequences commonly used during qualification that, literally, cannot be applied during real world engagements.

The simple fact is that most existing firearms qualifications are not representative, in any meaningful way, of job-related skills. Furthermore, training that is predominantly intended to prepare trainees for performance on these qualifications does little to prepare them for actual job performance in these mission areas. Using standard training methodology, qualification and job-related firearms skills are physically, tactically, neurologically and physiologically almost unrelated to one another.

With respect to this core issue that has long been facing the firearms and tactical training industry, **NURO™** *fundamentally changes the game*. Job-related performance potential evaluations (qualification) should involve measurement of the performance capability of the same physical, neurological and physiological mechanisms that are involved in job performance. **NURO™** is the first tool that facilitates doing this cost-effectively at scale. It allows these critical neurological functions to be always incorporated with, and connected to, the physical application of firearms and other use of force skills...

Introducing The World's Most Capable Tactical Training System

SHOOTING SYSTEM
Interactive Real Response Trainer

Do You Think While You Shoot?

Simulation-Based Training and Qualification. Empirical Standards. Situational Awareness. Vision. Decision Making. Policy. Judgment. De-Escalation. Mobility. Tactics. Patrol. SWAT. Military. CQB. Self-Defense. Concealed Carry. Indoor. Outdoor. All Weather. Live-fire. Dryfire. Simunition®/UTM®. Airsoft. Role Players.

Can you afford not to?

Now Accepting Pre-Orders

LASER LIGHT. AVOID DIRECT EYE EXPOSURE
CLASS 3R LASER PRODUCT
510-530nm and 630-645nm <5mW
EN/IEC 60825-1 2014

www.buildingshooters.com/nuro

Made in the USA
Monee, IL
30 October 2021